GREENSIDE

A Tale of Lakeland Miners

by the same author

FORCE CRAG: the History of a Lakeland Mine

G R E E N S I D E

A Tale of Lakeland Miners

by Ian Tyler

RED EARTH PUBLICATIONS
1992

ISBN 0 9512946 4 4

Published by RED EARTH PUBLICATIONS
7 Silver Street, Marton, Ulverston, Cumbria
LA12 0NQ

Printed in Great Britain by
SMITH SETTLE
Ilkley Road, Otley, West Yorkshire, LS21 3JP

Contents

Acknowledgement

MY THANKS go to numerous people and organisations who have assisted in compiling the details for this book.

The staff of the Kendal and Carlisle Record Offices for their help and assistance. The Kendal Office for allowing the reproduction of the account book and share certificate (WDB/35).

Mr J Borlase for his valuable information; Mr J Lawson of Beamish Museum Archive for his assistance and permission to use various photographs and documents; Mrs Grenfell of Carlisle for permission to use her photographs; Sir George Grenfell-Baines for information; Mr R Dawes for information; Mr D Sharman for his photograph; Mr A Findlay for his photographic skills; Mr M Carter for his photographic skills; Mr I Hebson for his photographs; Mr A Blezard for his photograph; Mr W Allison for permission to use his photographs; Mr P Blezard for permission to use his photographs; Mr A Lewis for permission to use his photographs; Mr Richard Bird for the use of his photographs; Mr N Thompson of Carlisle for his geological expertise; Mrs Beaty of Holme Farm Patterdale for information; Mrs O Walker of Liverpool for information; Mr L Greenbank of Kendal for information; Mr S Murphy of Staveley for information; Mrs O Allison of Cummersdale for information; Mr E Brown for graphics; Caldbeck Mining Museum Archive for use of photographs; the Wolf Safety Lamp Company Ltd, Sheffield; Trustees of the Lowther Estates for permission to use their archive; Dalemain Estates for permission to use their archive material; Eimco Great Britain Ltd, Gateshead, for information; Mr P Davies of the Lake District National Park for his cooperation.

Without the assistance and help of the following members of the Mines of Lakeland Exploration Society, many of the underground photographs and certain information would not have been obtained: Warren Allison, Alastair Lings, Martin Carter, Edd Brown, Ian Hebson, Peter Blezard, Paul Witheridge, Alen McFadzean, Doug Sharman, Julian Sharman; likewise members of CATMHS: Dave Blundell, Anne Danson and numerous others who have helped.

Special mention must be made of the miners whose patience in recalling details from so many years ago helped to give a personal insight to the history: Mr T Hope of Penrith, Mr W Burnett of Glenridding, Mr A Lewis of Burneside, Mr E Pool of Glenridding, Mr W Hicks of Glenridding, Mr E Flitters of Glenridding, Mr H Richards of Penrith.

Finally and perhaps most importantly my thanks go to my wife Jean Tyler for her tireless energy in illustrating and editing the text, and for her help in compiling the manuscript.

Foreword

HAVING recovered from the pain and the pleasure of writing my first book, it seemed a natural step forward to attempt a second, for after all I was now a "writer", tucked away on a shelf in the British Library somewhere between Tolstoy and Shakespeare. I had chosen to write about Force Crag for my first book because it had the distinction of being the last working mine in the Lake District. This time I decided on the one which had been the biggest and most successful lead mine, Greenside in the valley of Patterdale.

Sadly the New Coledale Mining Company has vacated the site of Force Crag Mine, which will now never work again, its closure due not to any shortage of mineral but a serious collapse within Zero Level. It is hoped that the landowners see fit to leave the mill and remaining buildings as a lasting memorial to the miners and do not have it demolished as so many other sites have been.

In an attempt to make people more aware of our Lake District mining heritage and with the aim of protecting these sites, a new mining society has been formed, Mines of Lakeland Exploration Society. The base for the society, fittingly enough, is the Caldbeck Mining Museum which displays much of the county's mining memorabilia.

Within the following pages I have endeavoured to tell the story of the toil, hardship, humour, ingenuity and courage of the men who worked the hard green rock for the silvery galena at Greenside Mine. Not to be forgotten either is the historical past of the other lesser known mines in the Patterdale valley which had their own stories to tell.

In the course of my research I have had the privilege and pleasure of meeting and talking with some of the miners, and so it is to the men of Greenside that this book is dedicated.

IAN TYLER, October 1992

1 Beginnings

Looking down the winding gorge of Kirkstone Pass towards Patterdale, encircled by the towering, rugged crags of Red Screes to the west and the silver acres of Caudale Moor scree to the east, few people realise they are entering one of the oldest and most prolifically mined areas of the Lake District. Sadly, all of the old mines are now closed and will no more ring with the noise and clatter of the toiling miners. But their story will be told as we unfurl the history of this fascinating area.

For many years the only easy way into the valley was via Penrith and Pooley Bridge in the north, or through Matterdale. The Kirkstone road as we know it is relatively new, and the only way south was over the tops by the ancient pony trade routes which crossed Scandale Pass and led onwards to Ambleside and beyond, or over Kirkstone by way of the pony track to Troutbeck and Kendal. Both these routes are now popular walkers' paths, but in those bygone days it was the hawkers and pedlars who struggled along them, heavily laden with their wares, to the next market.

Patterdale and the Ullswater watershed is intersected by various valleys, all of which have been mined. The minerals won were barytes, iron, zinc, copper, and most of all lead which had, in some of the veins exploited, as large a silver content as 15 ounces to 35 ounces per ton of ore. This high percentage of precious metal meant enormous success for some of the mines in the valley, success never envisaged even in their owners' wildest dreams.

Glenridding, Patterdale — these are just place names to the many hundreds of tourists who pass through the beautiful Ullswater valley. Some stay a short while, perhaps boarding a steamer to go for a sail on the lake, and from this floating platform they are able to enjoy the grandeur of some of the most spectacular views Lakeland has to offer. The contours of a mountainous landscape change little in thousands of years, unless disturbed by a freak act of nature, or sometimes, maybe, by man.

Above Glenridding village, up the Glenridding valley, are to be found the remains of Lakeland's largest lead mine, with a history spanning many years from around 1690 until its closure in 1962. It is true to say that had it not been for this industrial interlude, little or nothing would be here in the shape of houses and residential community. Certainly the whole of this village and the surrounding area owes its existence and prosperity to the mine. Many people were almost totally reliant on the work it provided.

Patterdale owes its name to St Patrick (Patricus), who had the misfortune to be shipwrecked on the Duddon Sands AD 540. On his overland journey he came upon the hamlet and baptised the locals at a well near the present church. Here a wooden church was built and named St Patrickdale, this being mentioned in records dated 1348. The original church was rebuilt in 1600, and again in 1852. Within the confines of its walls, many a miner was to marry, have his children baptised and finally be laid peacefully to his rest.

Some time after the invasion of the Normans, large areas of land were given to the conquerors. The manors of Barton and Patterdale were given to Ivo de Talebois, brother to the Earl of Anjou. Later, the lands of Deepdale and Hayeswater Gill were given by William de Lancaster to his son Gilbert in 1184. A further bequest was made of the Manor of Hartsop, which included Kirkstone, Caudale, Dove Crag and Angle Tarn. All this was for a payment of one shilling at Pentecost, and a further 10 pence on the feast of St Martin. The Lancaster family resided at Hartsop Hall, and here during the year of 1463, a great marriage took place when Mabel, the only daughter of the Lancasters, joined in wedlock with Hugh de Lowther, and as part of the dowry Hugh was given the Manor of Hartsop.

This was once a quiet backwater consisting of a few small holdings, the biggest settlement being Low Hartsop in the southern end of the valley. The area was not without its violent intrusions, and indeed this was the case in 1648 when John Mounsey, who was lord of the manor of Patterdale, and who resided at Patterdale Hall, was drawn to arms to defend the village against a band of invading Scots. The local people led by Mounsey lay in wait high above the road, at Stybarrow Crag, where the invaders had to pass directly below with only the lake shore on the other side of them. At the given signal the ambush commenced with large rocks being heaved down onto the unsuspecting intruders, and amongst the crack of pistol and musket fire the Scots were put to flight. The locals, overjoyed with their victory, bestowed the title of "Kings" on the Mounsey family, who ever after were known as the Kings of Patterdale and who lived in the "Palace". The family were great entrepreneurs and provided much work in the valley over many years.

In 1825 the Mounseys sold their estate to the Marshalls, who came from Cookridge in Yorkshire. They pulled the old residence down and the grand old hall now lying within its delightful, elevated garden, and encompassed by its 800 acres, is evidence of the wealth of the newly arrived family. William Marshall soon settled and created much work in the valley, securing many of the paths, roads and bridges which are still enjoyed by walkers today.

In such a small community, many of the families carried on their trade and traditions for generations, and many of the descendants of the original settlers still live in the dale. Some of the better known families are the Dobsons, Pattinsons, Places, Grisedales, Mattinsons, Oglethorpes, Lancasters, Gibsons and Pools.

The date that mining actually started is not known for sure; however, the location is certain to be that on the southern side of Stybarrow Dodd at OD 2,770 feet. The original workings were by the ancients, who first worked the lead vein on the surface. Here the vein was exposed in the ravines, no doubt originally by water. Further down the fell where the ground cover is greater, there is much evidence of "hushing" to expose the vein, though this method of mining was unlawful in England just after 1800. Consequently, as the vein was further exposed, it was worked by a series of adits.

The eastern spur of the fell is known as Greenside, for the rolling green turf of the summit area, which eventually gave the mine its name. Such a gentle, pastoral name gives no indication of the rugged, deep ravines which gash this side of the mountain. Erosion by snow, ice and water crashing down the crags, and gouging out the earth over many thousands of years, has laid bare the Greenside Lode and its various cross veins. The elevation of these early workings is between 1,850 and 2,300 feet, making them some of the highest workings in the Lake District.

Historically, it would be marvellous to establish a link with the German miners who came to Britain during the reign of Elizabeth I to stimulate the mining industry with their superior knowledge and experience, but no connections can be found. The Germans were very thorough in their administration and documentation, and available records do not mention this site.

The main interest of the German mining concern, the Mines Royal Company, was in copper ore, whereas the main Greenside Lode is lead bearing. There is, however, in the very top section of the mine, at OD 2,250 feet, a large cross-vein consisting of barytes with some copper exposed, and this would probably not have gone unnoticed by the German miners. They were prospecting in the vicinity, as the records verify that B Fechtenbach was given 10/- for finding a new vein at Thirlspot in 1574, and also at Fornside where working had commenced ten years earlier in 1564. Both these veins are just two-and-a-half to three miles west of the Greenside lode, but if any work was done here by the Germans, the evidence has been lost in the mists of time.

It seems inconceivable that these adventurers did not know about the Greenside lead lode, and of course had it been copper, it would most certainly have been extensively exploited.

Some of the earliest positive evidence of mine working in the valley can be found in the Patterdale church records, which record a burial at the church on the 25th of November 1713, of one "Margaret Vaize, a washer of lead ore". A further burial recorded at Patterdale is that of a miner, George Smedley, who hailed from Derbyshire.

In a report made in 1787, Patterdale was described as an area where the inhabitants eked a living on the land as farmers and labourers.

This type of work was all that was available to them for survival in a rustic environment. However, this idyllic setting was soon to be disrupted by the advent of the miners.

Around the year 1750 a report stated: "these fellows who are in general the most abandonde and wicked and prolifigate part of mankind, no sooner settled here, than they immeidiatly began to propigate there vices among the innocent and unsuspecting inhabitents. The farmers listen incensed at the stories told by the miners of places they had never seen or even heard of. Many were introduced to drink and games of chance. There daughters, lured by promises were seduced, those who were married were tragically at the end of the mining boom, left to fend for themselves in poverty and shame". Not a very chivalrous start to our history.

Times changed, and once again tranquillity returned to the dale. A later report, dated 1794, recalls a weary traveller stopping at a small inn, which could well have been the White Horse at Township, known later as Patterdale, as few buildings existed at Glenridding. Here was found a homely atmosphere of simple good cheer, where the landlord apologised for the small selection of liquor. He explained that spirituous liquor had not been sold there at all until 1787, and wine had not found its way to the local inhabitants, apart from on one or two special occasions such as a christening or marriage.

One of the earliest leases was granted by the Duke of Norfolk in 1799 to a miner called Thompson to work in Glencoynedale, and it may have been this gentleman and his colleagues who were responsible for the surface pits and trials at the head of the valley. These old workings were known as Duke's Sump, maybe so called because of the title of the lessor. There is evidence of hand-drilling at this site, therefore ruling out the ancient workings of the German miners, whose methods predated the introduction of boring holes for explosive charges.

Another lease granted around this time was to a miner called Henry Howard, who was working with a mining agent, J Errington from Alston, but unfortunately nothing is known of their exploits.

Despite all the speculation as to when the upper Greenside lode was first worked, the fact remains that a considerable amount of ground had been excavated by 1800. During the earliest period of working all the material would have been won by hand-picking directly on the vein. This would create open works, which would eventually reach such a depth that short crosscuts would need to be driven from the west to drain the stoped-out areas. It is possible that these very early trials were probably taken out prior to the use of hand-driving using blackpowder.

To prevent these ancient levels from being flooded, the natural fall of water off the fell was diverted slightly to the west, where it collects in Mosedale Burn Tarn. During this period all the ore removed would be dressed and washed entirely by hand. A hard, cold job this, using the freezing mountain stream water, the rough ore cutting into the hands

of the ore-washers and the buckers as they dressed the ore, separating the hard quartz from the lead, with no gloves for protection.

This was a tough job at any time and bad enough on a warm summer's day, but at over 2,000 feet, without shelter and with Helvellyn but a short distance away, seeming always to attract the vilest of weather, it must have been a hellish, soul-destroying task.

At the end of a day the poorly-clad miners, female ore-washers and the youngsters would start at least a three-mile walk back home, some taking the trod to Dockray and Matterdale, others the path down to Patterdale and Hartsop. With them they would take the empty powder casks and blunted drill steels to be sharpened by the blacksmith at Rattlebeck, ready for the morrow.

During this period ore, once sorted, was then taken by pack mule to the smelter at Stoneycroft near Keswick. The ore was carried in leather bags or stout wicker baskets, each mule carrying between one and two hundredweight. The journey from the mine would start with the long gradual haul up over Sticks Pass, down into St John's in the Vale and into Keswick, around the western shores of Derwentwater and so to the smelter.

It is interesting to note that the original mining excavations at Greenside are referred to on many of the ordnance survey maps as quarries, which of course is incorrect. The three uppermost craters are the results of mining practice and were caused by the ground cover giving way over many years and falling into the massive stopes worked out beneath.

2 The Adventurers

The four men looked up towards the rolling green top of Greenside mountain, their gaze resting upon the long abandoned lead workings, now quiet and still. Little evidence remained of the old men's labours, apart from a few heaps of spoil, the odd stone wall, and the remains of a shelter and cobbled dressing floor. As they looked directly ahead they could just make out the small dark adit entrances in the eastern side of the ravine. None of the men spoke, wondering if this wild and derelict place could possibly make the fortunes which they sought. The icy wind cutting across their faces, heads bowed against the weather, they started back down the track, the horses carefully picking their way down to Glenridding.

From here they took the more sheltered valley road back to Penrith and the warmth of the Angel Inn, a few hours ride. It had been a long day and on arriving back, the four ate a hearty meal, eagerly waited on by Mary Westmorland, the landlord's wife. The men discussed this new venture long into the night. Finally, the four agreed to become partners, Joseph Westmorland, the landlord of the Angel Inn, Penrith, Thomas Cant, grocer of Penrith, Mr John Rickerby, gentleman farmer of Burrell Green, Great Salkeld, and Mr Curl, a businessman from Broes, Dumfriesshire. The adventure they were about to embark upon was going to affect the valley and its inhabitants for the next 140 years. The date was around 1818, and the parlour of the Angel Inn was, for many years to come, to be a regular meeting place for the men as they made their plans.

Initially it was agreed that they would formally apply to the Duke of Norfolk at Greystoke Castle for a take-note to work and assess the value of the mine. This was normal practice, and after proving that the mineral deposit was worth mining they would form a proper company with shareholders and convert the take-note to a full working lease. Unfortunately, before the whole of the transaction could be drawn up, the Duke of Norfolk died, and due to the size of the estate and complexity of his will, it was some time before the negotiations could be completed. Finally, a lease was signed by Henry Howard Esq in the year of 1820.

During the initial years some preparatory work was done and the lead vein showed great promise. As a result, a small company was formed to push the venture forward with a much greater commitment, and in the year 1822, the Greenside Mining Company was formed.

It may be noticed that when the company was floated, certain of the original adventurers did not continue as shareholders. However, it is a well documented fact that shareholders in mining concerns tended to flit from company to company in the hope of striking a bonanza. The missing shareholders were Mr Curl, Mr Rickerby and, most surprisingly, Joseph Westmorland although, it must be mentioned, he was not in good health and was to die in the near future. It was decided that a full time agent should be employed to dispose of the ore and the man engaged was John Errington from Alston.

Original Members of the Greenside Mining Co 1822

Name	Occupation	Residence	Shares
Mr Thomas Cant	Grocer	Penrith	10
Mr M Crosby		Kendal	8
Mr Roberts			8
Mr Gibson		Cumrew	8
Mr Spedding			8
Mr George Remington	Farmer	Penrith	2
Mr Jackson			2
Mr W Harrison			2
Mr William Marshall	Gentleman	Patterdale	2
Mr Isaac Whitwell	Grocer	Kendal	1
Mrs Mary Westmorland	Publican	Penrith	1
Mr Thomas Parker		Carlisle	1
Mr Tom Cant	Grocer	Penrith	3

The next few years were spent setting up the operation, which entailed clearing out the old levels and securing the stone arching at the entrances. The old men had driven at least four levels and the indications were that the lode had improved with depth. The uppermost tunnel on Greenside was the Top Level which had been driven about fifty yards. The Middle Level was fifty feet below and had been driven eighty-five yards, and the High Level had been the drawing level for the old men. This had established the vein to a depth of 150 feet. The ground cover here had been shallow and the bedrock was exposed by a series of hushes and trenches which are still visible today. This had made the early mining comparatively easy. Unfortunately, the glacial moraine below this point is extensive and covers the bedrock by many feet.

Some 270 feet below the High Level are the remains of an adit which is not mentioned or recorded on any plan or document, but which could be the first High Horse Level. Its heading is north, was driven directly through the moraine, and was commenced as a stone-arched level. The well-grassed over tip reveals a quantity of galena, proving that the early miners had actually made contact with solid ground and the main Greenside lode. The overburden of the moraine would have allowed copious amounts of water to stream in from the surface and would have caused the ancient miners great difficulty over the initial drive of thirty fathoms to the solid ground. All the timber supports and cut stone for the arching would have been brought by packhorse from the valley below and progress would have been painfully slow.

Near to the entrance is the ruin of a fair-sized building, no doubt the smithy and office. A large natural plateau, ideal for dressing, lies just a few yards to the south and nearby is a substantial stream providing water for washing. This had been a courageous and adventurous drive and its inevitable collapse spelt the end of the old men's early endeavours to strike deep into the depths of the Greenside lode.

The new company no doubt considered clearing out this level, but it would have been well below the other workings even at this time. The 270 foot distance in height from the High Level was somewhat ambitious, the usual being about sixty feet, and would have to be proved first. The area was ideal for the new mill site though, and this was earmarked for the future.

The company appraised the situation and decided that a substantial level should be driven from near the base of the slope. It was decided to put in a drive just below an old trial, at OD 1,910 feet. This is now known as Gillgowar's Level, named after miner Jason Gillgowar who struck the bargain for himself and his gang to drive the tunnel. The entrance and probably the full distance up to the bedrock would be stone-arched. The distance to the vein was only seventy-five yards and was soon blasted and dug out. How the blue lead would have glistened in the flickering candle-light as Jason and his muckers got down to the job of taking out the rich ore. The small plateau at the level entrance would provide an area for hand sorting until a proper mill could be constructed.

In preparation for this, a substantial leat nearly half-a-mile long was cut from near the head of Sticks Gill to bring plenty of water for washing the ore. It was decided that the levels should be stoped out upwards, using Gillgowar's as the main drawing level. To ensure the safety of the mine, whilst attempting to stabilise the ground cover, it was decided to backfill the stoped-out areas with waste. This was sound mining practice as not only did it support the area but it also meant that unwanted rock did not have to be trammed hundreds of yards out to day.

Lead ore was now being raised, and on the 15th of April 1825 the first recorded parcel of lead was sold for £120.15s.5d. The royalty paid to the lord of the manor at 1/9th was £13.8s.4d. This was just the beginning, and the vein was proving all that had been hoped for. But it was patently obvious that the hand-dressing methods were ineffective and a proper dressing floor would have to be constructed to handle the quantity of ore now visible in the stopes.

To further vindicate the anticipated success of the mine, the company decided to undertake the cutting of a new lower level. This was to be driven as a crosscut at OD 1,810 feet, the distance to the vein being about eighty yards. This again was a stone-arched level and was cut to a good size, 6ft 6ins high by 4ft 6ins wide, and fitted with elm rails topped with iron. This was to become known as the High Horse Level and was to serve the company for many years to come.

On either side of the adit a miners' bothy and office were built, whilst not far away, to the south, the new dressing floor would have some of the most up-to-date equipment including stamps, crushing rollers, hand-operated box jigs, trommel screens and buddles. The whole operation would be powered by waterwheels, the water being brought initially by leat from Mosedale Burn and Sticks Gill some half-a-mile to the west. The main leat actually came from the foot of Rowantree Crag, and this allowed the residue water to continue its natural course freely into the burn which had, at that period, not been dammed. Much of the leat had been cut a few years earlier to supply water for the dressing floor near Gillgowar's Level. This was extended the final few hundred yards to where the new mill was to be sited. Further leats were also cut in the peaty saddle of Sheffield Pike, which in turn fed into the main water leat. Two small reservoirs were also built to retain sufficient supplies in drought periods.

Some of the miners would need accommodation from Monday to Saturday, so a stone bothy was built, measuring 30 feet by 25 feet, on the north side of the portal. It had a stone-flagged floor, slate roof, and three out buildings. Here the miners who walked from the distant villages of Dockray, Matterdale, Penrith town, Pooley Bridge, Legburthwaite, Grasmere and Threlkeld would stay over. Most were God-fearing men returning home late on the Saturday and going to church on Sunday. They would return in the early hours of Monday morning with food for the week carried in a "palaise", a sort of long pillow-slip made of cotton which hung around the neck. The week's supplies would perhaps include some cheese, bread, onions and maybe some home-cured ham, the only luxury being tobacco for the clay pipes which were smoked at the time. The journey had frequently to be made in awful weather conditions, but no work meant no pay, so off they set, groups of men from each village, brothers, fathers, sons, their clogs clattering on the cobbles, bidding goodbye to wives and families.

At the bothy a young boy or disabled miner would be paid a few coppers to clean and keep the place tidy, and in winter to ensure that a fire was kept going to dry the men's clothing which hung from the rafters to dry. The conditions in these huts afforded little privacy, the bunks were two tier and it was not unusual for more than one miner to share the same bed. Hot food was apparently something of a luxury, perhaps tea or maybe the odd pot of broth in winter. Lighting was by either tallow or oil lamp.

It needs little effort to imagine what it must have been like, the sickly aroma from the tallow candles mingling with the sweat of forty to fifty men who had just worked a twelve-hour day in muck and muddy water. The local miners from Patterdale, Hartsop and Glenridding would no doubt travel daily to avoid this.

On the other side of the portal was built an office and smithy, measuring 45 feet by 15 feet, where the mine manager and surveyor would chart the progress of the operation and assay the ore. Here many of the bargains were agreed for the long-term contracts, and the daily hiring and firing and general distribution of the workforce took place. The blacksmith was of course a skilled and important man, and at a mine of this size there would be work for more than one. Sharpening and making drill steels was just one of his skills, and many a miner owed the size if his pay packet to the ability of the smith to put a keen edge on the jumpers. Regular tasks included shoeing horses and repairing carts and machinery.

The scale of the mining operation was now growing well beyond its initial humble beginnings, and the lead already raised provided some cash flow and cemented the foundation for growth. The planned expansion would require more money and plenty of it, so the board decided to float the company.

The Greenside Mining Company was floated in 1827. The lease was with Henry Howard of Greystoke Castle, dated the 31st of May 1827, to run for twenty-one years at a royalty payment of 1/9th of the value of the ore. The area covered by this lease ran approximately from the summit of Raise, north to the summit of Stybarrow Dodd, east to the summit of Greenside, south-east to Sheffield Pike, south to Glenridding, and back up to Raise.

Shareholders of Greenside Mining Company 1827

Mr Thomas Cant	Grocer	Penrith
Mr John Cant	Grocer	Penrith
Mr William Marshall	Landowner	Patterdale
Mrs Mary Westmorland	Innkeeper	Penrith
Mr Isaac Whitwell	Grocer	Kendal
Mr George Rimmington	Gentleman	Penrith
Mr Thomas Parker	Businessman	Carlisle
Mr John Rickerby	Farmer	Great Salkeld
Mr Jonathon Buckley	Cheese factor	Liverpool
Mr John Crosby		Kendal
Mr Thomas Harrison		Kendal
Mr Matthew Harrison		Kendal
Mr Lancelot Harrison		Kendal
Mrs Margaret Harrison		Kendal
Mr Daniel Hartley		Penrith

These were the men and women who generated the spirit of the new adventure, and their initial investment was to prove most rewarding. Within ten years, a sixty-fourth share worth £100 in 1827, was to soar in value to no less than £1,000!

During these early years, however, all was not plain sailing as the price of lead tended to fluctuate. At one time, around 1824, the price fell to just over £11 a ton, an unsatisfactory situation for a young and developing company.

Some of the early pioneers suffered ill health, and one of the original partnership, Mr Westmorland, fell ill and died quite suddenly. His widow, herself a shareholder, sold his shares to Mr George Rimmington of Tynefield, Penrith. Shortly after the formation of the new company Mr Cant, the grocer from Penrith, was taking the waters at Shap Wells when he unfortunately collapsed and died. Neither of the men were destined to see the fruits of their speculation.

The mine buildings and installation of the mill machinery were progressing well, and a good road had been constructed from the small hamlet of Glenridding. At the crossing of Swart Beck a superb stone-arched bridge was built, a testimony to the craftsmanship of the day, for every ton of lead and every ounce of silver from the mine was carted over it for 135 years. The new road certainly helped with the transport of the heavy equipment up to the mill site. Iron tubs, castings, rollers, rails, waterwheel sections and numerous other things were all brought up it.

Coincidentally, over in Nenthead, the recently refurbished smelt mill was now operational (1825). The reason for the modification of this already busy mill was to incorporate a new method of separating the silver from the lead during the smelting process. This had greatly increased the silver returns, and the method was to become known as Pattinson's Process, named after its inventor, Mr Hugh Lee Pattinson, a Tyneside chemist. The Greenside management were quick to decide to send their ore to Nenthead and so benefit from the extra returns. This meant the ten mile journey by mule trains to the Stoneycroft smelter would be made no more.

The lead was now taken by horse and cart to Penrith where it was transferred to a pack-mule train, each mule carrying about one hundredweight. The carriage costs were much more but the enhanced silver returns easily recouped these extra amounts. The earliest transactions were as follows:

Date	**Silver**			**Lead**			**Royalty**		
	£	s	d	£	s	d	£	s	d
1825	642	03	05	1202	01	07	198	15	01
1826	219	04	10	612	03	02	138	11	04
1827									
1828	239	10	11	518	00	01	126	05	02
Total	1100	19	2	2332	04	10	463	11	07

The royalty paid appears to have fluctuated over this period and was charged at either 1/9th or 1/6th.

During the early period, the population of Glenridding and Patterdale was very scant, existing mainly as a few isolated farms and homesteads. The overall population for the area in 1801 was only 269, dwelling in sixty-two homesteads. This had been increasing steadily as the mine opened up and the operation attracted miners and their families from all over the country. Miners have always been nomadic, moving where there was work available, and Greenside was certainly the place now. They lived frugally in lodgings where they could get them, many single men staying at the hostels as there was not enough accommodation locally. Some local men also, from places like Penrith, Threlkeld, and Grasmere, turned from working the land to the new challenge of lead ore mining. This, they were soon to find, was just as hard a life.

Most miners provided their own candles, which were purchased from the mine shop, and many also provided their own tools and bought their own blackpowder. This was because the miners were not employed directly by the mining company, and jobs were won as contracts or bargains. A group of perhaps four or five men would have a spokesman who would negotiate a deal with the company, perhaps to drive a level at so many shillings per fathom, or build a stone-lined water leat for a hundred yards. The men having agreed the bargain between themselves would then sign the agreement with the company. On completion of the work the men would have the money logged, and if it was a contract which would take months, the men would be given goods and credit at the company mine shop which sold most general items.

The technique of driving levels at this time was by hand-drilling. This entailed usually two men working in pairs, one holding the drill steel whilst the other man struck the steel from behind with a sledge hammer. The man holding the steel would keep turning it slowly, and the dust and bits were either blown out with the mouth or scraped out with a special tool called a shot-hole scraper. Drill steels varied in size both in diameter and length. Generally the shorter steels were used in harder rock and measured approximately twelve to eighteen inches. Longer ones of twenty-four to thirty-six inches were used in the softer rock.

Once the hole was cleaned out and free from damp, the blackpowder for charging would be placed into the hole. Once in the hole the powder was then tamped down with a wooden charging stick although in the very early days iron bars were used with sometimes disastrous consequences. Wood or brass was used to prevent sparks and pre-ignition. The top part of the hole was then stemmed with clay to ensure the powder charge was sealed into the confined space, and would so produce maximum blasting power. A hole was then pierced through the clay with a brass charging pricker, and a straw or slow burning fuse then inserted and lit by the miner's candle. The men then withdrew rapidly to allow the gunpowder to do its work.

To drive a level of about six feet high and four feet wide, the miners would drill up to twenty or more holes to obtain forward progress of perhaps two or three feet, this process being repeated time and time again. The distances driven varied with the type of rock encountered.

Gunpowder was first used in actual underground blasting in the Ecton mines in Staffordshire around the 1670s. Prior to this the ground was worked by fire setting, plug and feather or pick. One of the most famous gunpowder factories in Cumbria was sited at Elterwater in Langdale and was known as the Elterwater Gunpowder Company, founded in 1824 and working until 1928. During its time it supplied hundreds of tons of powder to the Cumbrian mines.

To Make Saltpetre Circa 1561

Fyrst black earth the blacker the better. The next is Urine, namely of those persons whiche drink either wyne or strong bears. Then Dong specially of those horses, which be fed with ootes, and always be kept in the stables. The fourth is Lyme made of plaster of Parys. The Lyme whiche is made of Oyster Shellis is the best. Mix the heap and turn over a period of months, untill sufficient salts have formed. Boil down the resulting liquid and allow to cool so that the crystals of Saltpetre Precipitate.

Work underground was carried out by the flickering, yellow light of tallow candles made mostly from animal fats. The sickly smell created by the melting fat, particularly at the end of a blind heading, would have made conditions pretty obnoxious to say the least. In many cases the miners had to purchase their own candles, and consequently they were used very frugally. It is a well known fact that many of the men worked in teams of two or more. Fathers and sons, brothers and friends, bonded together in a life time of hard endeavour, their reputation for skill with the hammer and drill steel often going before them. Some were known by their names for their hard work and proficiency in driving, and it was these men who earned the most money.

The usual clothing worn at this time, and indeed up to the Great War, would be clogs filled with straw, this straw often being used as primitive fuses. Thick woollen trousers, tied at the ankle or knee with a piece of twine known as a "yorker" to stop the dust or the occasional rat running up the leg. A woollen shirt with a leather or cloth waistcoat would be worn to cover the top of the body, but in the deeper recesses of the mine the men would work stripped to the waist. Around the neck a kerchief was worn, and to protect the head, a cloth cap. During earlier times, miners wore basically the same type of clothing with the exception that the jacket may have been leather with a long back.

The pay days during this time took place twice a year at the Angel Inn, Penrith, and here Mr Errington paid the men their dues. This

was a very important day and the miners would make the most of it, some taking their wives or sweethearts, some going by boat to Pooley, others by cart and the less fortunate on foot. Mr Cant's grocery shop in Penrith thrived on these days as no doubt did the Angel Inn. The bargain system seemed to work fairly well, the men getting the price they wanted by an agreement which was binding, once signed for, both to them and the company.

By 1830 the mine was considerably developed, the new High Horse Level well under Gillgowar's Level and ground being stoped out through to the old higher workings. In High Horse Level the miners had commenced underhand stoping and, happily, the vein was proving to be as rich as ever. No doubt based on these findings the directors decided to build their own smelt mill. The carting charges to Alston, coupled with the smelting costs, had been a constant drain on resources; in fact, the cost for smelting and carting alone for 1826 was £118.12s.4d. A site was selected at the confluence of the Swart and Glenridding becks. The site was perfect with plenty of water power, and well below the present mill and the main adit, and with a good height clearance for the flue to ensure a strong updraught.

It was now certain that the vein still proved good in the upper ground, and with this in mind the company were determined to capitalise on their good fortune. A major decision was taken by the board to ensure the future of the company for years to come. Based on the ore being removed from the understoping in the High Horse Level, it was proposed to drive a main drainage adit thirty-five fathoms below, thus opening all the ground between. This new level would become known eventually as the Low Horse Level and would take many years to drive. It also meant an increase in the total number of men working at the mine which was then around eighty.

In the High Horse, the stoping and general exploration were going well, with ore being drawn and taken to the new mill. Any waste material not being used for back-filling was dumped at the adit mouth to make a causeway to the mill. The sheer volume of waste being created started to fill the tarn itself. As this was a natural and regular source of water, another dumping site had to be found. It was decided to commence tipping on the south-eastern side of the tarn and culvert the water underneath the tip. Below this point, near the dressing floor, a small but secure dam was built. This ensured that water was retained at the higher level before it plunged down the foaming cataracts of Swart Beck. It was felt that the venture was secure for many years to come. However, more changes were afoot.

3 The Greenside Mining Co 1835

It was 1835 and Mr George Head-Head, a banker of Rickerby Park, Carlisle, and another local gentleman, Mr Joseph Morland, had bought into the company and, along with several others, formed a new board. Some of these positions were created by the deaths of members of the earlier board. The cash investment and the fact that a leading banker was now involved gave the operation greater credibility. The company expanded its production to somewhere in the region of 600 tons of dressed lead ore per year.

Not all the activity was underground, by any means. The engineers had to plan the Low Horse Level entrance and site carefully, for here Swart Beck follows a savage ravine and is totally impassable when in spate. The point was selected at OD 1,600 feet, actually inside the ravine, but before the level could commence a sturdy bridge of iron and wood had to be built to span the gorge. This was strong enough to carry the weight of a pony and at least two or three loaded iron tubs across to the proposed new dressing floor. The first loads of barren rock from the driving of the level were used to create a small elongated area above the ravine for the smithy and other buildings. More rubble was used for the foundations of the new ore receiving area.

The next problem to be overcome was the disposal of the vast quantity of waste from the driving of the tunnel. It was realised by the engineers that thousands more tons were to follow and so it was decided that a further bridge would be built to recross Swart Beck. Again the engineers were put to the test as here, some 100 feet above the water, they spanned the gorge for the second time.

Much of the new mill was constructed of local stone, and the size of some of the foundation blocks would do a cathedral proud. Some were hand drilled and pinned together with huge iron rods. All were placed with precision and care by the masons who realised what the vibration of heavy machinery could do, not to mention the rigours of Lakeland weather. Much of this area still remains today as it was then, and stands as a monument to the men who toiled in its construction.

Meanwhile, the actual driving of the tunnel itself was painfully slow through very hard rock. It was blasted forward virtually twenty-four hours a day, the men working three shifts. Conditions at the face must have been horrific, only one small ventilation rise having been put up to surface. This was just ten yards inside, and was driven to effect a water blast (a crude ventilating device), and which no doubt later powered a waterwheel for a fan to pump air to the face. No further ventilation was available until the miners made connection with the upper workings, some five-hundred yards in from day. But this was not to be for many years.

The temperature underground is about fifty-five degrees, but the humidity, with the sickly aroma from the tallow candles and the perspiring workforce, and mixed with the fumes from the discharged gunpowder, created almost unbearable conditions for the men. Each yard of work they had bargained for squarely, but they could never be sure what lay beyond each barrier of rock as they went slowly and painfully forward.

Below the Low Horse was another hive of activity as work started on the new smelter. The undertaking of these projects had increased the overall workforce considerably, particularly on the surface. This was not only due to the increased production, which required more people in the mill area, but many jobs were now being created in the construction of the smelter which required waterwheels for the bellows and leats.

Most miners, as we have said, were God-fearing men and sober, unless there was something special to celebrate such as the twice-yearly pay day. On the 7th of March 1835 the men set off to the Angel Inn in Penrith to collect their pay, which as usual was dispensed by Mr Errington, the mine agent. On receipt of their money some miners bought supplies whilst others went for a mug of ale, either staying at the Angel or going to the nearby Miner's Arms. The influx of the miners into the small town created a lot of activity, and many pedlars were there to sell their wares. The miners would later return to their homes ready for work on Monday.

Two of the men who had returned to the dale, Joseph Bainbridge and a man named Greenwell, both originally from Alston, were drinking in the White Lion in Township (now Patterdale village), having a final fling before Monday. An argument started and a brawl ensued with some of the locals, and the two men were ejected by the landlord Mr Sissons. The night was dark as the two men headed back up to the mine, and fearing they might be attacked they decided to cut a stick to defend themselves. Whilst in the process of doing this they heard someone approaching, but in the gloom they could not see who it was. Greenwell, knife in hand, lunged at the oncoming figure and the blade entered the unfortunate person's abdomen, mortally wounding him. Realising that their fears had been mistaken, Greenwell and Bainbridge carried the man up Chitty Puss Lane to Jane Chapman's house, but here he died. The man was identified as Thomas Grisedale, of Hartsop Hall, aged twenty-seven, who was just walking home and had not been involved in the original affray.

This terrible incident resulted in the two men being arrested for murder. They were held locally, prior to being transferred to Appleby Gaol the following week, but the weather was atrocious, the ground deep in snow and the police had great difficulty in transporting the prisoners through the huge drifts. Mrs Westmorland of the Angel Inn knew both the men and took young Greenwell a new shirt, as the one he was wearing was still bloodstained from the incident. She was a compassionate woman and felt he should look more respectable for the trial.

It was nearly a year later when the trial was brought before the Assize Court at Appleby. Greenwell stood in the dock, a mere twenty years of age, of quiet disposition and of previous exemplary character. Alongside him stood Joseph Bainbridge who, by contrast, was a big man, rough and ready yet gentle in manner unless in drink, when he could become abusive. The law at that time was hard, justice was swift and Greenwell was sentenced to death. It was unusual in those days for a condemned man to be reprieved, but young Greenwell was lucky and his sentence was commuted to penal servitude for life. Bainbridge was acquitted; he did not return to the Greenside mine but went to work in the Durham coalfield where he gained a reputation for being one of the finest miners in the district.

It was this incident which brought about a major change in the company policy, and from 1836 it was decided that no further pay days would be held in Penrith. Instead the men would be paid at the mine. By now the total number of miners working at Greenside was at least a hundred, and although some accommodation was available at the mine for the men, there was little spare housing in the village for their families who were employed as ore-washers and labourers. This was becoming a serious problem as both Patterdale and Glenridding were nothing more than hamlets, and the influx of the workforce would continue as the mining operations grew.

It was not only people who were brought to the dale — many extra horses were brought in requiring fields and stabling. Extra land and stable facilities were rented from Rev John Thompson and Henry Howard Esq in July 1838, costing the princely sum of 5/- per year.

The development of the mine continued as did the construction of the smelter. The road up to the Low Horse was examined by the engineers and was considered inadequate for the transportation of ore from the new mill at the level mouth. Sorted ore was still being processed at the mill at High Horse and brought down the old track, a laborious job by either mule or cart.

The engineers considered that an inclined tramway should be built, and down this the dressed ore would be more efficiently transferred to the smelter. This project would take a considerable time, but the company were far-sighted and were prepared to invest in the future. The years rolled by and the money came in, and the production figures (lead produced and sold) were still steady.

The mine was progressing at quite a pace, both underground and on the surface, with new buildings being erected and many contracts offered. For example, a bargain dated the 23rd of June 1843 was agreed between the company and miners William Richardson, Joseph Richardson, William Vipond and Lanty Merryman to stope the roof for £3.16s.6d per fathom. On the 4th of August 1843 a bargain was struck for the carting of wood by George Stalker, of Patterdale, who agreed to take wood from the Lake Road to the smelter for 1d per foot, or from the smelter up to the Low Horse or High Horse levels or the miners' shop at 3d per hundredweight. The shop was built to supply victuals to the men and was sited at the south side of Swart Beck near the new crushing mill.

The blacksmiths were very important and respected men and, true craftsmen that they were, they passed on their knowledge and expertise to their sons so that the tradition would be carried on. Thomas Watson was the blacksmith at this time, and his contract for three months from the 23rd of June until the 23rd of September 1843 was as follows: new jumpers 4d; copper prickers 2/6d; hammers and picks 5d; steeling old jumpers 3d; shoeing horses 5d; separate wage for drill steel sharpening, 13s per week.

Times indeed were becoming prosperous and the directors faired well. The dividend paid during the period ending the 7th of March 1845 was as follows:

Shareholder	**£**	**s**	**d**
Mr G Head-Head	900	0	0
Mr G Rimmington	400	0	0
Mr W Jameson	400	0	0
Mr W Whitwell	300	0	0
Mr H Harrison	250	0	0
Mr Stafford	150	0	0
Mr T Hartley	100	0	0
Mr T Parker	100	0	0
Mr Crosbie	100	0	0
Mr T Harrison	100	0	0
Mr W Hunter	100	0	0
Mr Cant	100	0	0
Mr R Lawson	50	0	0
Mr R Jameson	50	0	0
Mr D Hartley	50	0	0
Mr J Thompson	50	0	0
Total Dividend Paid	3,200	0	0

The prosperity of the mine was now proved and a further dividend of the same amount which had been paid in March was issued on the 25th of June 1845.

The lead from Greenside had a very high silver content and this contributed greatly to the financial success of the mine. Nearly all the silver was dispatched to the Royal Mint in London, where it was made it into the coin of the realm. During 1845 the average monthly payment for silver was £375. The quarterly wage bill at the mine during the same year was £2,004 though this did not include the directors' wages.

The company were generous enough to allow the men to sub their wages and this appears to have been a regular occurrence as a total of £762.5s.0d is recorded as having been drawn by the men.

During 1847 and 1848 the production figures were down, but according to the accounts a great deal of rebuilding work was being done particularly at the smelter. The firm of John Blaylock of Carlisle supplied a large amount of fireclay, whilst the brick firm of G H Ramsay was supplying fire bricks in large numbers. The firm of Crowley Millington supplied ladles for the smelter at a cost of £5. It was probable that the smelter was being modified for the new Hugh Lee Pattinson process of separation, for the accounts state that Mr Pattinson was paid a royalty that year for the first time of £37.0s.6d.

The price of lead varied between £16 and £18.5s.0d per ton, whilst silver was bringing 5/4d per ounce. During 1847 the major customer was the Royal Mint who took 9,458 ounces of silver for which it paid £2,538.9s.7d. The biggest client for lead was the firm of Walker Parker Ltd of Newcastle upon Tyne who took 4,000 pigs, although Newton Keates Co of Liverpool was not far behind with 3,000 pigs. A firm called Lock & Co took 1,500 pigs, the Kendal Union Gas Co 100 pigs, and a William Marshall took 100 pigs. The pigs weighed between 112 and 120 pounds and the total number sold during this year was 8,700. The amount realised by the Greenside company was £8,637.7s.9d which made the total income for lead and silver £11,175.17s.4d.

Year	Tons	Cwt	Qtr
1838	764	6	1
1839	452	17	1
1840	792	3	0
1841	642	9	0
1842	622	7	3
1843	738	19	1
1844	758	17	1
1845	1339	6	3
1846	1157	11	2
1847	902	6	0
1848	819	4	3
1849	1287	6	0
1850	1223	8	3
1851	1232	18	1
1852	948	19	2

As the mine was being driven steadily forward, gunpowder was a regular purchase and came from either Wakefield & Bainbridge, Curtis Harvey & Co, or D Huddleston & Co, Elterwater, later to become the Elterwater Gunpowder Company. The usual amount paid was £49.17s.6d for the delivery of a hundred barrels per month. The 100lb kegs were brought by cart and deposited in the powder magazine which is situated just below the mine. When the delivery was being made a red flag was hauled up the flag pole to make people aware of the possible danger. After delivery, the powder was taken underground in smaller wooden kegs weighing around 20lbs. Candles also were another regular item on the company's shopping list, and were supplied by Crosfield & Co from whom they purchased about £40 worth per month.

Much heavy plant machinery was brought in from Messrs Haydon Bridge Iron Co for heavy castings, no less than eleven transactions being made, indicating further major renovations. Inevitably, when modifications on this scale were made, a large amount of scrap metal had to be disposed of and this was purchased on a regular basis by the Carlisle firm of Cowan and Sheldon. This firm, incidentally, became one of the biggest crane manufacturers in the world until they closed down in the late 1980s.

4 Down to Lucy

George Head-Head was a member of the Society of Friends and in his efforts to sway the community to his beliefs had built a meeting house at Grisedale Bridge as an alternative to the Wesleyan methodist chapel. Shortly after this, the generosity of the Greenside company extended to St Patrick's when a silver communion plate was given at the Easter Offering of 1849.

During this time much of the ore was transported by cart to Penrith, and from here it was dispatched by rail to its various destinations, London, Newcastle, Liverpool and other places. Lead was a very important commodity in those days and when in pigs was easily disposed of. In spite of the weight, thieves managed to steal some eighty pigs (four tons) from Carlisle while they were in transit.

The Low Horse Level was being driven at this time by T Swindal, J W Patty, J W Holliday and T Holliday for £9.10s.0d per fathom. They were driving in very hard rock and this is reflected in the price. A prospecting drive west off the Low Horse, known as Marshall's Level, was into much softer rock and miners J Watson and J Little were driving this for only £4.5s.0d per fathom. Here the face was some 150 yards from where it left the Low Horse and was in a bad place to work. There was no ventilation and the level meandered its miserable way through barren ground with little or no ore to be found. There would certainly be no bonanza stopes here. A few hundred yards further on along the Low Horse Level a trial was made on the Johnston Ross Vein but this was also proving relatively poor.

Eventually the Low Horse came underneath the upper workings proper, and here a sump known as Whim Sump — sunk in the floor of the High Horse Level — had nearly made the long awaited connection. This would provide the vital ore pass from the upper workings, create some ventilation for the miners and also provide a second way out. The High Horse mill was still working, but as soon as the Whim Sump connected, the upper mill would become virtually redundant. The ore receiving area at Low Horse Level horizon was now complete with smithy, offices, bouse teams, a waterwheel for primary crushing and two huge ore bins at the incline head ready to transfer the ore down to the smelter, although, as yet, the incline to the smelt mill was not finished.

During 1851 a census took place, and it is interesting to note that the population had increased during the past fifty years by 416 to 685. With all this development work going on it should not be forgotten that the intense job of mining ore was still continuing, with virtually all the lead being won in the high workings. The High Horse was now well developed and was some four-hundred yards in to the face. All the levels above were interconnected with a series of manways and rises, and the stopes were now right through from the High Horse to Gillgowar's and into the very old High Level.

The board of directors, always eager for further development, now came up with a new plan. This was to be a massive undertaking and, if executed, would mean that many of the board members would not see its completion. It was planned to capitalise on the new smelter site and the incline for transportation to the much lower position. The directors were so convinced of the richness of the vein, that they envisaged years of work in the Low Horse, not to mention the stopes still to be cleared in the High Horse workings.

It was therefor decided that a complete new mill should be built just above the smelter (this site is now totally obliterated by the slimes and waste from the modern workings), and that a new horse level, the Lucy Tongue Level, should be driven at OD 1,081 feet to cut the vein, which was about a mile away. The magnitude of this proposal must have created many heated arguments, for the cost was going to be enormous and the project would take many years.

It was one Joseph Cowper Cain, an eminent mining engineer from Alston, who endorsed the company's forward planning and, based on his surveys, the driving of the great Lucy Tongue Level eventually got under way. As one can imagine, an undertaking of this magnitude was fraught with problems. These began at an early date when Mr Cain required to survey the Low Horse in order to ensure an accurate plan of the workings. He was assured by the board that he would be given every assistance and the equipment to carry out the task. On reporting to the manager's office to collect the surveying equipment and commence the project, he was aghast at the inferior and out of date instruments he was offered. He insisted that new equipment be obtained immediately from Troughton & Sons of London, but this took weeks to arrive. Finally, Mr Cain was ready to commence the survey but then realised, however, that work could only start from midnight onwards as the dust from blasting and the general filth within the working sections of the mine prevented accurate reading of the instruments. These numerous delays resulted in the survey being months behind schedule, which in turn prevented work starting on the new level.

Perhaps the most significant aspect of the survey was the indication that a new engine shaft in the Low Horse Level was down a total of thirty-six fathoms. This had been put down in extremely hard rock and the sole of the shaft was now only forty-two fathoms above the intended Lucy Level. It was Mr Cain's recommendation that three main levels be taken off the shaft. These would be designated main sub levels, and would contain hoppers and be railed out. The first to be driven was the 36 Fathom Level.

By the 6th of May 1853, discussions between Mr Cain and the directors were well advanced, and the proposals by Mr Cain for the Lucy Level were discussed in detail.

It was estimated that six miners driving in the soft ground of the Lucy Tongue Vein would drive sixty fathoms a year. Mr Cain had speculated from his survey that there would be 374 fathoms of this type of ground to cut and this would take about six years. There would then be a further 319 fathoms of hard rock, and with the best men driving they could hope to cut only twenty-four fathoms a year, thus adding a further thirteen years to the drive. The full length of the drive would be 693 fathoms and would take nineteen years to complete, a drive that fathers would start and sons would finish.

The huge dog-leg drive which followed the Lucy Tongue Vein was not the shortest route considered by Mr Cain. An alternative and more direct drive was possible, but this was through much harder rock although at 556 fathoms it was some 137 fathoms shorter. Due to the nature of the harder ground, however, the level would take some 23 years to complete. Mr Cain was well aware of the rock strata, as the miners had encountered this type of rock in the sinking of the engine shaft. It blunted the drill steels and required much powder to blast it.

The financial commitment to the project was already immense and the extra cost for this alternative route was far in excess of the budget. Although the shorter, harder route would cost more in powder and take longer, the level would be safer with little or no outlay on timber, whilst the longer route would be driven more quickly, because of the softer ground, and any weak areas could be timbered or stone-arched. Mr Cain favoured the longer drive and, assuming his estimates were correct, it would mean this area of the mine could be opened more quickly. Eventually the board agreed and decided that the longer level would be the most prudent. The Lucy Tongue Level was to be one of the longest crosscuts in Lake District mining history and would be cut to a good size of nearly 7 feet high and 5 feet wide.

The position for the adit entrance was selected with great care and strategically placed just above the smelter. Just to the east across Swart Beck was a large barren area ideal for the thousands of tons of waste which could be dumped without interfering with the smelter operation and the planned mill. It is hard now to imagine the chaos existing at the time, as the mining activities were spread all over the hillside from the smelter right up to the High Horse workings.

To accommodate the influx of workmen a new bothy was completed and this was to be the home for hundreds of miners. It was sited just above and to the west of the smelter. The cost for the 41 foot by 18 foot building was £10.10s.0d. The specification was that the walls had to be 2 feet to 3 feet thick, fire places were to be set in each gable end, the walls were to be mortar rendered inside and the roof slated. The gentlemen who undertook this contract were Thomas Boustead, Robert Pattison and William Fleming, and the contract was completed during 1855.

Nearby at the smelter, which operated twenty-four hours a day, a contract was signed by Cuthbert Hetherington for one year, to wash all the black slag from the slag hearth. Smelter Joseph Glenwright, with his partner and a labourer, guaranteed to work not less than three days per week, otherwise the Greenside company could fine them 2/6d a day. R Bainbridge and J Potts also worked the smelter and were paid 14/- for a week's work. J Lancaster who worked nights until 4:00 am at the calcining furnace earned 3/- per night. Labourers at the smelter were E Pool, J Kitchen, J Stanger, A Weir, J Robinson and J Archibald. This was a hot and dirty place to work, and the fumes were dangerous. The men smelting were all craftsmen, but many of the labourers came just for the money. Few stuck it for more than a few weeks though, and in some cases a couple of days was enough at the furnace for 1/3d per shift.

During this period of activity the mine was working to capacity. The upper workings were still being stoped, and some ore was still being crushed and sorted at the High Horse mill, although the underground connection between the Low Horse and the High Horse had been effected. This had allowed hundreds of tons of ore to fall directly from the upper areas to be drawn from the Low Horse where it was crushed and sorted at the Low Level mill and fed down to the flames of the smelter furnace. Here the men toiled day and night in appalling conditions where only the fittest survived. It was said that on a clear day one could see the smoke billowing from the chimney stack high on Stang End.

On the 22nd of February 1855, Mr J C Cain met two of the Greenside directors, Mr Harrison and Mr Rimmington, to discuss the progress and development of the mine. They met at the mine office, and after exchanging courtesies, the three men proceeded to the Lucy adit just a short distance away. The level adit was now well secured, and across Swart Beck a substantial bridge had been constructed and laid with track to transport the tons of waste material.

Underground, the directors, who were experienced mining men, were quick to notice that much water was breaking through into the level even though it had only been driven sixty fathoms. Mr Cain assured them that this presented little problem, as the level would be well secured at this point, and that some seepage from the beck above was to be expected. However, the directors were still concerned and suggested that Swart Beck be diverted higher up the gill which would alleviate the problem and also give more ground cover. This could be done by building a retaining wall of stone and waste from the adjacent Low Horse Level tip. This was agreed and the work commenced right away.

The western side of the leat was blasted out of the solid rock, and massive foundation stones weighing many hundredweights were dragged into place to be hand-drilled by the masons and pinned to the bedrock with irons. The wall was built some 6 to 8 feet high and of a similar width, and work on this was carried out over the next few years, whilst the miners toiled beneath. Most of this stonework is still standing and working as efficiently as it was designed to all those years ago, a credit to those fine craftsmen. Changing the normal route of the beck also allowed more water to be drawn off at a number of leats at a later date.

The two directors, happier in their minds, commenced the walk up the hill to the Low Horse Level. Here Mr Cain complimented them on the fine structures of the Low Horse crushing mill and ore receiving area. The self-acting incline was still in the process of construction. On arriving at the Low Horse adit horizon, the men could survey the whole panorama of this tremendous venture.

Below them lay the smelter, its huge flue snaking up the hillside, hissing and puffing smoke and fumes. Clanking tubs rolled in and out of the levels and the never-ending crunching sound of the crushers mingled with the groan of the waterwheels as they strained under the endless deluge of water. Nearby, in the smithy, horses were being shod and steels sharpened in preparation for the next shift. Miners, joiners, labourers and horses were going to and fro about their work. It was a sight to gladden the eye. Certainly the prosperity of the mine seemed assured.

As the men discussed the day's affairs, the flat truck pulled by a pony arrived to take them into Low Horse Level. They pulled away from the smithy, and the men quietened as the truck creaked across the ravine and carried them inside the level. Here the flickering of the candles revealed little, and after four-hundred yards of seemingly endless tunnel, they arrived at the junction of the barren Marshall's Level. Continuing on they came to the Engine Shaft (later known as Willie Shaft) and here they alighted. The shaft top was clear and the men descended a few fathoms and entered a sub level running north.

Satisfied with their inspection, they returned to the Low Horse and then visited various headings, most showing good lead in the face. Thirty fathoms north of the Engine Shaft, a further shaft had been sunk and this connected back to the Engine Shaft at a lower horizon. They were advised that this area was ready to be stoped. Mr Cain commented on the large quantity of ore lying in the hoppers, and it was explained that this had not been drawn because of the awful weather they had been having recently, and that the drawing would be effected as soon as the weather improved. This had not, however, affected the mill output as the reserve ore bins were full. Well satisfied with the inspection the trio regained their transport and returned to day.

Underground, the serious business of pushing the ground forward was continued by a variety of teams: L Thompson and partners to drive at 115/- per fathom, J P Swindle and partners at 120/- per fathom, T Sibson and partners at 120/- per fathom, and J Thompson and partners at 85/- per fathom. The differences in the rates were due to the type of ground encountered and whether the men were engaged in stoping, driving or sinking.

All work was done on contract or bargain basis including cartage. For instance, Richard Metcalf and Joseph Crump agreed to transport all manner of goods from the 1st of April 1855, for two years. A special penalty clause was inserted in the contract regarding the transportation of coal, which was subject to pilferage by the locals. The clause read: "Should the coals from Penrith be short of weight on arrival at Greenside the loss to be made good by the carters". The rate of 5/10d per ton was agreed and the account would be paid every quarter. Another carter's bargain was struck by John Robinson of Gillside Farm, but this time it was for the transportation of lead ore from the High Horse workings and crushing mill down to the smelter. The price agreed was 1/3d per ton, provided all the ore was bagged. Lime brought from Fluscoe lime kilns near Penrith cost 5/- per load delivered to the smelt mill as required and the bargain was for two years from the 14th of May 1855.

During 1856 activity was centred on the driving of the Lucy Level which so far was in a total of 130 fathoms. Mr Cain, after visiting the face, was disappointed at the low yardage driven for the operation had fallen behind schedule. He recommended that, in order to get back on line to meet the estimated completion date, the level should be driven forward twenty-four hours a day. No doubt some urgency was lent to the situation by the recent increase in the price of lead, which had been brought about by the extra demand for munitions for the Crimea War. During this period lead was £23.12s.0d per ton, an increase of some £6.2s.0d over the previous two years. The post war period was also going to reflect a sound price and in actual fact Great Britain, during the ten year period of 1851 to 1860, smelted more lead than any other country, producing some 740,500 tons which was nearly 41% of the world market.

By the end of 1856 many changes had taken place at the mine. The whole of the upper workings had been virtually worked out and much of the ground between the High Horse and Low Horse levels had been stoped away. The height of vertical ground removed was in the region of 700 feet, and the miners had taken nearly 23,700 tons of lead and silver valued at £545. They had driven upwards of three miles of levels, all of which had been achieved by hand-driving and manual physical effort.

Gunpowder for the operations was still being supplied by the Elterwater Gunpowder Co and the fuses supplied by the Garrigill Fuse Co, owned by John Hall of Windy Hall. Candles were supplied by Dawson and Bushy. The largest single expenditure for 1856 was a new smelter chimney, constructed by the John Pattinson Co for the price of £984.8s.0d. When fully completed the chimney was nearly a mile long and finished at OD 2,250 feet. It was of stone-arched construction and lined with wood throughout most of its length. The stone came from quarries on Stang End and the Stang itself, where the flue eventually ended. The special fire bricks to line the lower part were supplied by a firm called G

H Ramsay.

On the 7th of April 1857, J C Cain again visited the mine to see how the Lucy Level was progressing and to assess the ore reserves in the old high workings. He was also concerned at the amount of effort which had been expended on driving a forty-four fathom prospecting drive off the High Horse Level. This had been driven east and was totally barren. So Mr Cain, accompanied by Mr Addison, Mr Barron and Mr Rickerby of the Greenside Mining Company, made their way up to the high workings.

It had been four years since Mr Cain had been in this part of the mine, and as they entered the High Horse Level (known also as the High Way Gate Level), his trained eye noticed the poor condition that the level was in. It was apparent that, due to lack of use, as the workings had deepened and the subsequent connection been made with Low Horse Level below, there had been little or no maintenance carried out. Timbering and packing were down, and obviously the deterioration had been accelerated due to poor ground cover allowing surface water to pour in through the glacial moraine.

The miners had pursued the remains of the vein upwards and had cleared the very top-most workings above Gillgowar's Level. In fact they had broken through the bedrock, and consequently the ground cover had actually broken up and was in danger of falling in on the men working hundreds of feet below. This greatly concerned Mr Cain who pointed out this danger in no uncertain terms. Frustratingly, some of the ore body remained, and it was decided that a plan should be worked out for the careful removal of some of the remaining mineral.

The workings between the High Horse and Low Horse were investigated and again the workings in the sole of the High Horse were found to be in poor condition. Mr Cain observed that the vein had been worked more vigorously in the southern end rather than in the northern headings as had been originally agreed. On arriving at the Engine Shaft on the Low Horse, however, he found that excellent headway had been made. The headframe and sections had been set in the shaft and it was equipped for a man-riding cage — and tubs could be taken down to the levels below, saving much manpower.

The richest ore-producing areas at this time were the stope just north of the Engine Shaft, where the vein was producing both above and below Low Level, and another vein called Harrison's, also producing good ore, but because of the very hard ground here the men were having great difficulty in taking the lead.

When a good strike was made the miners were put to work stoping the ore out, dangerous work due to the chance of rock falling from above. It was generally found that bargains for working the stopes were less than for sinking shafts and driving levels. In many instances stope work was assisted by gravity as the ore could simply be crowbarred or pickaxed out of the roof, and although dangerous it was not classed as skilled. It was especially hazardous when the miners were working high in the stopes, and certainly after charges had been set off, leaving huge hanging blocks which would crush a man flat. Ever present also were the noxious fumes caused by the discharge of the explosives.

Miners working the roofs at this time were Fenwick White for 32/-, Adam Craig for 25/-, J R Fothergill for 22/-, Edward Cherry for 38/-, John Shields for 24/-, and Thomas Gibson for 30/-, all contracted for one year and paid per fathom. This was hard, dirty and frightening work, the men standing on wooden staging in huge voids, many feet up, with only the faint glimmer of the yellow tallow candles to see by. There was no protection and their work was accompanied by the din of the continuous crashing and banging of the ore tumbling into the hoppers below, and the dust and acrid fumes rising as the hard rocks ricocheted and splintered on impact. Fearful work indeed.

Satisfied with their inspection so far, the gentlemen proceeded outby and so down to the Lucy Tongue Level. Here there were more problems. On checking the driving heading Mr Cain discovered that the miners had accidentally driven off the Lucy Vein on a stringer, and they were now heading off forty degrees to the west. The problem was left with the Greenside men with the comment from Mr Cain that the compass must be used with more care in the future, and recommending that a regular set of plans be kept on site and that they be kept up to date.

Next the dressing floor was visited, where it was apparent that the method of transferring the ore from the crusher was inefficient, but because of the layout of the mill this was impractical to change. In view of the intended increase in production it was considered that the slime pits should be increased and extended into three ranges of sixteen pits. A water race would be constructed below to ensure any spillage would be contained.

The only other observation made by Mr Cain was that at Rake Head the mine road had a vicious "S" bend which he considered to be too sharp for a long timber wagon, and tortuous for the lead wagons to negotiate. This was to be taken out as soon as possible, but it seems this remark went unheeded as the bend is still there to this day. Apparently not all of Mr Cain's advice was taken.

5 Work Goes On

In the year 1860 Greenside Mine underwent yet another major transition, this time concerning the surface working and in particular the smelter. Hundreds of bricks and casks of fire clay were brought up to the mine for further modifications and extensive repairs. The main smelter chimney was originally shorter than its completed final length of about a mile. At its termination point the stack was nearly 1,200 feet above the smelter. The flue cut through the surface of the hill to a depth of about six feet and about four feet wide. Along its length there were numerous condenser boxes for collecting the residues, and also small junction points were built into the flue which, when opened, would assist the drawing of the fire.

Originally, when the flue trench was dug, it passed through many quartz veins, most of them barren, and as a result it acted as a huge surface exploration trench. There were, however, two small trials driven, one on the south side of Stang End breast. This level, the timber portal of which is now totally collapsed, was driven only a few yards, but galena is to be seen on the dump. Due south of here and across the gill there is an impressive stone-arched level extending just twenty-five yards to a blind heading with no mineral apparent.

During this period the mine was still producing lead and silver at a steady rate, and the shareholders enjoyed all the trappings of being involved in a successful company. Some of this success may have been attributed to the fact that the American Civil War had been raging from 1862 to 1864 and this had helped steady the lead price to £21.2s.0d. Records do not reveal that lead from Greenside went to America, although it is not impossible.

An important survey was carried out at the mine by John Gill of Leadhills, Scotland, beginning on the 3rd of April 1866 and continuing for a period of three days. The surveying party first of all entered the old High Level, 260 feet above High Horse, and here they entered an area called Shaw's Stope in the eastern workings — but little was left. An inspection in the western stope, however, showed ore still in the face. The ore drawn from this stope would pass down through to the High Horse via the Old Hopper.

The party then proceeded down to the High Horse Level and climbed down the ladders in Gillgowar's Sump where four miners were hard at work at a four-foot vein of lead. Nearby, a short trial had been driven off east in Gibson's High Sump but this showed little lead. From here they went to the site of the Big Crush, a huge collapse caused by the ground cover being undermined by the stopes. Surface water seeping in over many years had dislodged tons of material, much of it vein stuff, and this the miners were now in the process of working out. This was, in fact, the area which Mr Cain had considered so unstable many years before.

The magnitude of the Big Crush had to be seen to be believed. The slip had occurred back in 1862 when a gigantic piece of vein material, 60 feet thick by 170 feet long and 160 feet high, containing some 6,000 cubic feet of vein and weighing around 120,000 tons, became detached from the side wall of the High Level stope. Miraculously, this happened on a Sunday. When this monolith began its downward journey, ripping away all before it, dragging out the carefully-placed deads and floor pack of Gillgowar's Level, its massive bulk crashing through the High Horse and finally shattering into the narrow stopes of the Low Horse Level below, there were no casualties.

Providence had indeed been kind, for had the mine been working the carnage would have been truly appalling. The estimated value of the lead was £80,000 and it took four years to remove. Amazed at this unforgettable sight, the survey party continued on, descending through the High Stope, near the Irish Hopper Rise which was in poor condition. Near here, the vein was split into three headings, all of which showed little ore.

Descending a few more ladders they arrived at the Low Horse Level. This had been driven forward 100 feet that year and was now in a total of 401 fathoms. The group were impressed with this area of the mine, which was being worked vigorously, and a solid rib of galena, at least eighteen inches thick, was to be seen in many of the working areas.

After this they descended the ladders to the Engine Sump which was now down forty-nine fathoms. Sub levels were being worked north and south at the 36 Fathom and 48 Fathom levels. Returning back up to the Low Level the men proceeded out to day and walked down to the Lucy Level.

On entering the Lucy Tongue Level they inspected the wooden air trunking and waterblast, which had been set up to get air to the face to ease the discomfort of the miners. In the last two years the men had driven a total of 235 feet and there still remained at least 605 feet to the Greenside Vein.

On returning to day John Gill and his party visited the crushing mill. Here they saw the new waterwheel which had been installed, and which had greatly assisted in obtaining an increase of 144 tons on the output. They also found that work on the smelter chimney flue was complete and had been extended a final 540 feet. This would ensure that the smelter worked more efficiently. More residues could be retrieved from the flue to be re-smelted and the silver and lead recovered. It was as a result of this visit by Mr Gill that a complete section plan of Greenside Mine was accurately drawn up, giving much detailed information and indeed serving as the basis for many later plans.

At this time the mine manager was Mr Richard Grenfell who was later to become manager of the mines for the Patterdale Mining Co, formed by some of the shareholders of the Greenside company. Richard was a Cornishman, born in 1816 and married in Mousehole in 1842 to 16-year-old Mary Harvey. He attended the Cambourne School of Mines but then changed course and set up business in Mousehole as a general supplier. He later went into shipping and traded in Cuba and the Caribbean for four years. The loss of some of his ships created financial problems and so he returned to his original profession of mining.

Richard and his wife Mary, along with their nine children, came to the valley in 1864. Their long journey was by sea from Penzance to Bristol, thence by rail to Penrith, and on by carrier to Glenridding. Here they settled in Eagle Cottage where their tenth child, Hedley, was born soon afterwards. An eleventh child was born but did not survive and Mary also died soon afterwards. The family grave is in Patterdale churchyard and the stone bears the following inscription:

Sacred to the memory of Mary Grenfell
beloved wife of Richard Grenfell, Agent in the
Greenside Mine, aged 42, died February 1st 1868
"O husband dear my time is past
My love remained while life did last
And now I'm dead no sorrow take
But love the children for my sake"
Also to the memory of Richard Grenfell
who died June 11th 1875 aged 59
"Until the day dawns"

In July 1867 Greenside Mine acquired another mining agent, Thomas Raw, who came from the large lead mining area of Arkengarthdale in north Yorkshire. He was born in 1811 and had worked for at least eighteen years as a miner, learning his craft and eventually graduating to the position of manager at Surrender and Arkengarthdale mines, and later achieving the status of agent. He was actually managing the Old Gang Mine at the time of his move to Greenside and on arrival in Patterdale took up residence in Greenside Lodge.

However, his appointment with the company was somewhat short-lived, and various items of correspondence indicate that something went amiss. Certainly his eating habits seem to have been rather extravagant as an entry in the accounts for May 1870 reveals the cost of a dinner as £2.15s.0d — and this has been heavily underlined. During the second week of August 1871 Thomas left Greenside to return to his native Yorkshire and Blakethwaite Mine. An item of documentation at this time states that Thomas Raw left Greenside "with not quite clean hands". In spite of such a short stay it was long enough for him to have his name immortalised in the annals of Greenside, as one of the angled sumps cut down from Lucy Level bears his name.

The mine continued to expand and consequently there were many more purchases made. During 1867 rails were purchased from William Whitwell & Sons for £119.3s.8d. This would have been at a time when railing in mines was changing from the old wooden rail clad with iron strips to the new pressed steel rail, which of course was more durable. Another modern item on the shopping list was wire rope from R S Newell and Co, and various amounts were purchased. The wire rope would be for use in the hauling shafts, and for the inclined tramway.

Around this time the first sinking in the Lucy Level, the New Engine Sump (Lucy Shaft), was commenced. This coincided with the strange purchase listed in the accounts for steel jumpers, bought from the Coniston Mining Co, for which payment was made of £3.8s.11d. Possibly, these were required for the work on this shaft.

It was during this year also that the Keppel Cove earth dam was constructed. This was to collect and retain the water in Keppel Cove. When completed the pond area would be 6.2 acres and the water, at its deepest, would be some thirty to forty feet in depth. The dam was about 50 feet high and 150 yards across, the parapet being nearly 60 feet wide and the base approximately 270 feet in breadth. During the later stages of construction an adit tunnel 150 feet long was driven through the dam which carried an 18 inch cast iron sluice pipe for release of the water. The dam was built with an 18 inch spillway and was designed to have a five foot clearance at the parapet. It was officially examined and passed as safe by government inspector John Harle on its completion in 1871. The dam was required to feed water to the new mill being erected at the Lucy Level and smelter, and the leats to serve this operation are still clearly visible to this day.

The smaller, unfinished dam slightly higher up, in Brown Cove, was made just afterwards to serve as a feeder to Keppel Cove. It was possibly felt, however, that because of the capacity of water now stored and readily available due to the naturally high rainfall, further expense could not be justified and so the project was abandoned.

The mine manager Richard Grenfell who, incidentally, was being paid £8.8s.0d per month, was no doubt excited at his future prospects, now being in charge of the most progressive and successful mine in the area. The workforce was expanding steadily, and it was seen fit by the company to employ Dr Gilruth on a private retainer basis to look after the welfare of the employees.

In 1867 the Lucy Tongue Level was well in the Greenside Vein but the Engine Sump from the Low Horse Level had still not made a connection. Jewell's Sump, however, had broken through to the Lucy. The directors were delighted, for this now opened up the whole of the mine. For the toiling miners it meant a second way out and the luxury of fresh air, for many had toiled for years at the Lucy face, stifled with the dust and filth, the only ventilation being the trunked air which puffed feebly out and which gave little comfort.

The mine was now ready to be opened up entirely, the many years of exploratory driving and sinking to make the vital connection had been achieved. The investment had been worthwhile and now it was ready to be exploited. Consequently, the mine required a greater workforce and in order to accommodate the influx of miners and their families coming into the area, two rows of cottages were built by the mining company and rented to the miners at £1.5s.0d a month. Some of the earliest tenants are now household names in the Greenside history, names like Miles Shaw, John Millican, Lancelot Pattinson, Thomas Coward, Edward Pool, Robert Dargue, and William Nelson. No sooner were the cottages built than the eager tenants moved in. By 1869 nine cottages were occupied. By the end of the year the number had risen to twenty-one.

More cottages were built beside the mine road, those on High Rake and Low Rake, two rows on Stybarrow Terrace, and two rows called Halton Terrace in Glenridding. A further row was built at Seldom Seen in Glencoynedale. The local ale house had now come into being and was known by the locals for years as "the Jerry", so-called because the ale was served in pots, and was a favourite haunt for the thirsty miner for many years to come.

The bargain system was still the norm and as a result it is difficult to establish the total number of employees, as one man normally signed for his mates. But the following, taken from the records for 1870, will give some indication of numbers.

Workforce May 1870

Miners	91
Wagonmen	3
Breakman	1
Cagemen	2
Labourers	36
Masons	5
Joiners	16
Blacksmiths	7
Horsekeepers	3
Smelting	9
Roasting	9
Preparing	12
Calcining	2
Weighing	1
Shepherd	1
Washing floor	1
Carriage	2
Total	201

The Greenside Vein was pushed north both in the Low Horse and the Lucy Tongue levels, while in Glencoynedale a crosscut level was driven forward at OD 1,790 feet, just below and to the east of the old Duke's Sump workings. Here, in May 1870, William Jackson with three other miners, and George Bell with another three men, were working a bargain of £28.4s.8d.

This new crosscut was a bold attempt to test the main lode. This is a solitary adit, situated in high ground and miles from anywhere. The long unremitting climb to this lonely outpost is frustrating also, as height first gained has to be lost. The only compensation is the incredible view down Ullswater, though doubtless the miners did not consider this. In fact, a substantial path had been hacked out of the hill by the old men, which circumnavigates the head of Glencoynedale. This was the trod used when they had walked in from Dockray or Dowthwaite Head, and also was most likely used by the mules taking the ore to Penrith from the Old High Mill. And so once again it was put to good use.

Meanwhile, just a few hundred feet below, the northernmost end of the High Horse Level, which had been virtually barren, was now in the process of being bricked up to make a dam for underground water storage. Once the hundreds of yards of level had been sealed the reservoir of water contained within would be used to power a hydraulic double-cylinder engine which would give a pressure of 80lbs psi. This would be used for winding and pumping in the Willie Shaft (previously known as the Engine Sump and named Willie after the first brakeman). The water would be sent down the 200 feet of shafts through a twelve inch iron pipe, the vertical head giving it the required pressure.

The first man to be given the important job of brakeman on the winding shaft was Willie Glenwright. It was his responsibility to see that tubs being either raised or lowered were not overloaded, and that the men were standing in the cage properly and did not exceed the stated maximum number. The job paid £6 per month but Willie was sometimes unpopular, refusing to heed the cry of "Go on, Willie, you can squeeze one more in".

The men drawing the wagons from the shaft top were James Thexton and George Bell, who received £9.6s.11d between them — not much when you think of the hard graft, dragging the huge fully-laden iron tubs out of the cage till they had a rake of four so the horse could run them the mile out to day. They were sometimes envied though, as the shaft top was lit by a paraffin lamp, and the air was usually quite fresh — unless of course any of the ponies were badly trained.

Perhaps the most important work being carried on at this period was on surface where the new dressing floor foundations were being put in by a team of workmen under Mr Foulston, and a road was being constructed to the new mill and quarry. John Robinson and his partners had bargained to dig a waterwheel pit some 24 feet by 8 feet by 8 feet. The foundations of the new blacksmith's shop were also being laid. Leats were being dug by no less than eight gangs of men, who had been working on the newly-constructed dam at Keppel Cove. It was during this period that all of the mill operation was moved from High Horse. The Low Horse still did some basic crushing, but most of the ore would now come through Lucy Adit to the new mill.

A secondary smelter flue was in the last stages of completion, and although the flue was working it was still being finished off by John Brumwell and John Atkinson, and the final job of covering the flue was done by Isaac Peart.

By December 1871, underground, the miners were still pushing forward the Lucy Level which was now around 330 yards past the New Engine Sump, this having been sunk to a depth of forty fathoms. Here a level was driven off north and south, though some of the ground was poor and so was stone-arched by Daniel Allison to ensure its stability, whilst just above, at the shaft collar, joiner David Bellas and partner were replacing the shaft cage gates.

The Willie Shaft had been sunk a further twelve fathoms and was now down to the 48 Fathom Level horizon, and from here levels were being driven north and south. Indeed, it would be some time before the remaining thirty-four fathoms would be sunk and the shaft connected to the Lucy Level. There was actually a connection already between the Lucy Level and the Low Horse, a three stage sump called Jewell's Sump, the bottom part of which had been put upwards from the Lucy. Another connection was the Brook's Sump, just south of the Lucy Shaft and partially inclined. Much of the stoped work from the 36 Fathom and the 48 Fathom levels probably came this way, and Jewell's Sump most likely served as an adequate manway. This is probably why the Willie Shaft did not hole out until 1882.

High in Glencoynedale, the Glencoyne Adit was virtually completed and by 1873 the miners had driven a distance of 140 fathoms. It was driven north and south in the vein with some crosscuts but, alas, nothing of any value was found apart from a barytes vein. In the southern heading a 30 foot sump was put down on a more promising looking part of the vein, but this too was a worthless trial. No doubt this aggravated Edward Pool and his men who endured the slog up here day in and day out bringing the gunpowder and steels from the stores some 1,100 feet below.

However, their labours were not entirely in vain, for it was decided that the barren drive could in fact be put to good use. It seems that with little effort the High Horse Level could be connected by a 30 foot shaft. Then the barren section of the High Horse could be dammed at strategic points to create more water storage for the hydraulic system in the main workings below. Eventually the entrance to the Glencoyne adit was sealed with bricks and timber, and thousands of extra gallons of water were drawn into the level by a covered watercourse which was put in from the direction of the old Duke's Sump workings. This was then put down through a small shaft into the Glencoyne level.

The reservoir, once filled, was to be used for the second hydraulic engine which was being constructed at the top of Lucy Shaft, using the spent water from the Willie Shaft engine, the extra 480 foot head giving 212lbs psi. Water was also fed in from the Top Dam via the High Horse Level to ensure the capacity of the internal reservoir was maintained.

The Low Horse Level was now in a total of 670 fathoms, approximately 70 fathoms further north than the Glencoyne adit. Disappointingly, the final 150 fathoms of driving was barren, the last strike of ore worth stoping being almost directly below the sump Edward Pool and his men had sunk.

At this time there were several boys employed at Greenside Mine, many of them the sons and younger brothers of miners and other workers. The laws of the land were changing, and one such law referred to the employment of children. The Metalliferous Mines Regulation Act 1872 required that all children under the ages of twelve for boys and thirteen for girls working above or below ground be listed. During 1874 the law was revised, making the ages for boys, thirteen and girls, eighteen.

Young Employed at Greenside Mine, 1st of July 1874

John Armstrong	16	Glenridding
Joseph Brown	13	
John Cole	14	
Charles Jewell	13	
John Kilner	17	
William Kilner	15	
William Lake	16	
Richard Lake	14	
John Murray	13	
Edward Pool	15	
William Pattinson	13	
Robert Shaw	14	
John Wilson	14	
William Robinson	15	Keswick
Joseph Milburn	17	Patterdale
John Robinson	16	Vale Of St John
Joseph Robinson	14	
William Harrison	17	Matterdale
David Harrison	15	

It was the policy of the Greenside company to employ people of this age on the surface only. The Edward Pool mentioned above was presumably the son of the aforementioned miner of the same name, but sadly young Edward was not destined to live beyond his eighteenth year. Although exact details are not available it is recorded that Edward died accidentally whilst working as a wagon lad on one of the levels and it is most likely that he was crushed to death by a moving tub.

For the size of the mine there were not many accidents, but in the year of young Edward's death, 1877, there occurred a catastrophe caused entirely by the laws of nature. Near to the gill side by the smelter stood the silver refining house where all the silver from the mine was kept prior to being taken away to London to the Royal Mint. All was safely locked up one night when the men finished work and headed for home. The evening was still and heavy, and dark clouds unfolded over Catstycam mountain, thunder growling in the sultry air. The first big drops of rain began to fall, sending the miners hurrying quickly to their homes as the storm began in earnest.

The next morning dawned fair and clear, though the men noticed as they made their way up the road to work that the gill was a torrent of white water. At the mine, all was utter pandemonium, men running here and there and the bosses ordering everyone up to the upper mine area where evidently the Top Dam had burst. The silver refining house had been washed away by the storm and with it a 1,000 ounces of silver worth £300.

By today's standards this may seem a negligible amount of money, but when compared to the difference in wages for example, it becomes clear what an enormous loss this was. A miner at that time was earning £6 a month, or £72 a year. Today, a miner could expect to earn £250 a week, or £13,000 a year, not including bonuses. This puts the value of the silver lost at about £54,000.

It is a puzzle why the so-called Top Dam was the one that broke. It does not seem feasible that the hundreds of gallons of water released could have broken through the thousands of tons of spoil which were lying in its path and which would have acted as a second dam. None of these spoil heaps has been disturbed, and they extend too far to the south to have been by-passed by the water. One theory is that the spillway or safety valve allowed a large volume of water to pass through the sluices and back up behind the Old Dam at OD 1,750 feet. This is just below the old High Horse mill site and to the east of all the spoil heaps. When this dam burst (and it does appear to have done so) the torrent of water would have plunged directly into the confines of the ravine walls and, assisted by the cloudburst on that night, have raised the level of the beck sufficiently to have caused the damage. The story goes that the miners were panning the beck for weeks afterwards, but none of the silver was ever recovered.

Though the hard-working Greenside miners were never destined to make their fortunes in the dale, others went farther afield to try their luck. One ex-Greenside man who was lured away was Thomas Hutchinson, a part-time huntsman for John Crozier of Threlkeld and a man who, over the years, had many mining adventures of his own. Thomas had three sons: Charley who went to America, and Jim and John who, with a friend Ned Walsh and a Mr Hethorn of Melmerby Hall, went to the Australian gold fields. Mr Hethorn barely survived the long sea voyage and indeed died the day after arriving in the colony. It was, however, Ned Walsh who returned a wealthy man two years later with 360 ounces of gold.

Month by month the numbers of young Greenside employees grew. It was inevitable that some accidents would occur and that some would be serious and others fatal. Willie Jackson had been proudly blessed with four sons, all of whom had followed their father's profession in the mine. One day, two of the boys, Robert and John, were descending a manway, Robert going first. On reaching the level he shouted up that he was clear and John started down but slipped. Hearing his brother's fearful cry, Robert realised what was happening and braced himself to prevent John striking the floor of the level. The force of the impact on Robert was so great that his ribs were crushed into his intestines and he died of his injuries. John however, survived because of his brother's action.

On the 14th of May 1880 an important contract was awarded by the Greenside company to Warsop and Hill of Nottingham, inventors of one of the first compressed-air rock drills. The drills, weighing between 65 and 70lbs, were manufactured by Messrs Tangye Bros & Rake, Newcastle upon Tyne, and worked at a pressure of 16lbs psi. Warsop and Hill were awarded the contract because of the speed with which the driving could be accomplished using the new equipment. Initially the contract was to sink a shaft 75 fathoms deep by 12 feet by 9 feet, at a price of £48.10s.0d per fathom, and to drive 250 fathoms of level, 7 feet by 5 feet at £16.4s.0d per fathom, and a further 37 fathoms at £10.7s.0d. per fathom. The work was to be carried out three times faster using the machine drills than by traditional hand methods. Failure to maintain this would result in the price of the contract being reduced accordingly. This was providing the miners did not strike or the water fail.

If the contract times were maintained, Warsop and Hill were promised a second contract to sink another shaft 30 fathoms deep and 12 feet by 9 feet at £48.10s.0d per fathom. A further clause in the contract stated that the Greenside Mining Co would provide all labour, power, rails, props, wagons and kibbles, remove all rock debris from the working faces and keep the shafts drained. With the exception of the debris from the shafts, the kibbles would be loaded by Warsop's own men. The Greenside company were to pay Warsop and Hill every month. Some of the driving carried out was the crosscut between Smith's Shaft and Willie Shaft — this was to become known as Warsop's Crosscut.

The new air drills were the first to be used in the mine, and much greater speeds of driving were achieved.

By 1887 the Lucy Shaft was down about forty fathoms and still going, and the Lucy Level was well past Willie Shaft where, just to the east, was the site of the proposed New Engine Shaft, later be known as Smith's. Just to the north, nine partners led by William Hicks were energetically sinking a sump by hand through to the 40 Fathom Level, and for this they had bargained £14.0s.0d per fathom.

Due to the Lucy Shaft and Willie Shaft being serviced by hydraulic engines, another brakeman had been recently appointed. This was Edward Pool, and another engineman, John Hetherington, had also been trained as a relief. John and Edward were paid £4.8s.0d per month whilst William Glenwright was now paid £5.8s.0d as senior engineman.

Lighting was still predominantly by tallow candles and these were supplied by Thomas Smith, normally in boxes of two dozen costing

£1.15s.0d. The Elterwater Gunpowder Co continued its endless supply, QB powder at 32/- per barrel, and HF powder at 35/- per barrel (HF and QB are codes for the power of the powder).

It was rare for horses to be kept underground for days at a time as was the norm in coal mining, and for years the company rented Gillside Farm, where the horses were stabled when not underground. The rent for the farm was £120 a year. To supplement the food requirements of the mining community, sheep, geese, pigs, and cattle were kept, and some general produce grown for the mine shops.

During this time vast quantities of coal were used at the mine, supplied by Thompson and Son, and a load of 73 tons 4 cwt cost £48.16s.0d. Most would be for the smelter. Nearby, the new mill was well under construction and the Haydon Bridge Iron Co had just supplied one jaw breaker crusher, four sixteen-inch crushing rollers and twenty incline wagon wheels. This would imply that the incline was now almost ready for use. A new stone breaker from R Marsden was also supplied.

A significant change in blasting methods took place when, on the 5th of November 1887, Henry Kitchen Ltd supplied 150 pounds of dynamite for the price of £7.17s.6d and 350 pounds of blasting gelatine at £26.15s.6d. The Garrigill Fuse Co supplied a hundred coils of safety fuse at £3.6s.8d. This was no doubt for Warsop's operation.

The site of the New Engine Shaft had been prepared although the actual sinking did not commence until 1890. The coming decade was to be one of the most important in the history of Greenside Mine and, indeed, there were to be several firsts during this period.

The year of 1888 started badly at Greenside for on the 28th of February Robert Gibson Bell, of Hartsop, while returning to day after the shift, was killed instantly when a large section of the roof came down. Tragedy struck the mine again in the same year when William Clark was preparing the face for blasting. As he removed the pricker prior to inserting the fuse, he struck a fire and detonated the charge which killed him instantly. It was recorded in the local paper at that time that there had only been four fatal accidents at the mine in twenty-four years.

Towards the end of the year the mine was employing two-hundred people, and on the 28th of November the prosperity of the company was again reflected in its dividend, a £25 share returning a £50 dividend. The future for the mine seemed assured.

Richard Grenfell and his wife Mary came to Greenside in 1864. Their long journey was by sea from Penzance to Bristol, thence by rail to Penrith, and on by carrier to Glenridding. Here they settled in Eagle Cottage where their tenth child, Hedley, was born. An eleventh child was born but did not survive and Mary also died soon afterwards.

A team of miners boring a shothole by the time-honoured hand method. Note the candles fixed to the rock walls in balls of clay. Bottom: miners taking a smoke in the roof of a stope, steel jumpers heaped in a pile in the middle of the photo.

Miners working in the roof of a stope, boring shotholes with an Ingersol Rand boring machine.

Charging up the shotholes with dynamite. Protruding from one of the holes is a wooden charging stick, used for tapping down the explosives and packing the stemming material.

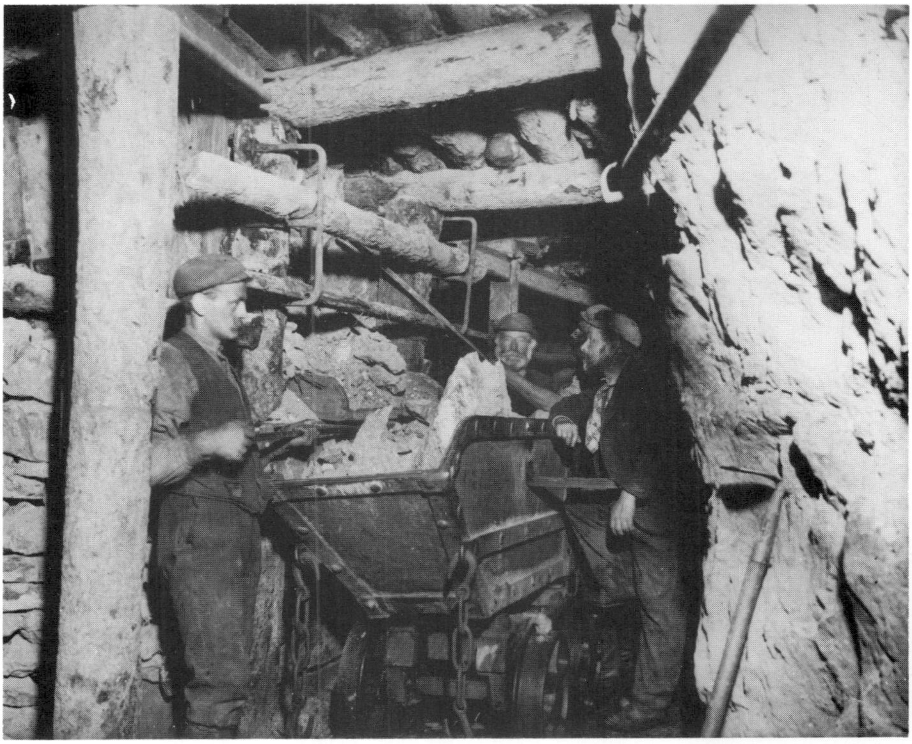

Miners filling a tub with ore from a hopper. The man wearing the tie (extreme right) is Captain W H Borlase.

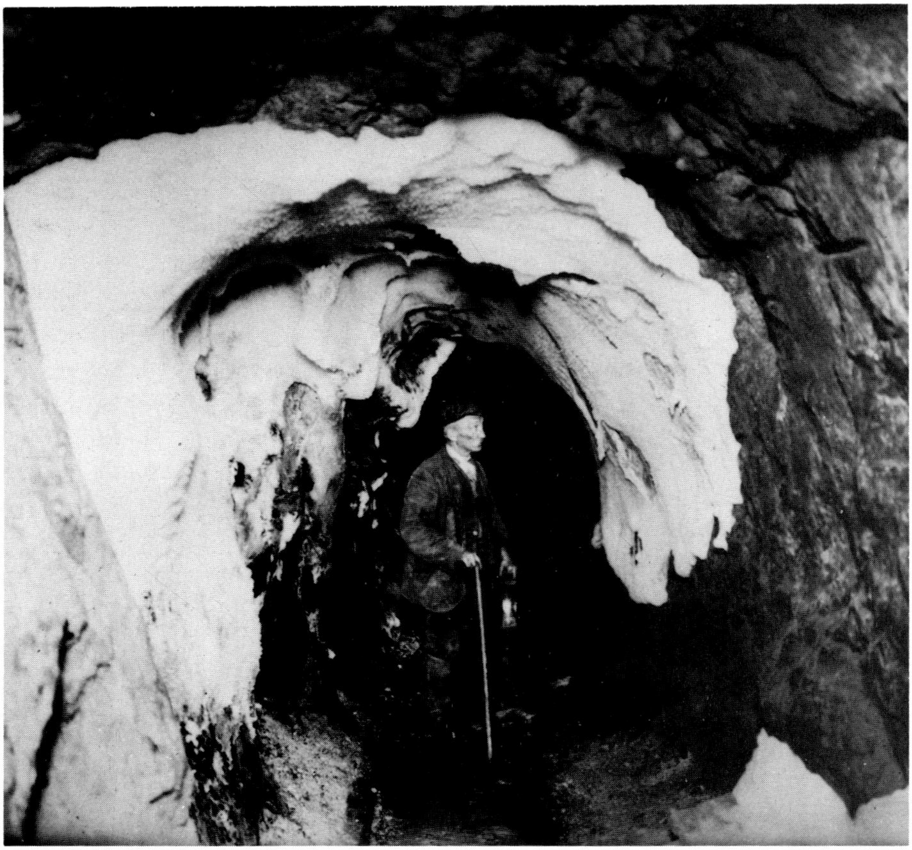

Deep down in Greenside Mine, fungus forms on the walls of a level.

Smith's Shaft top circa 1900. Smith's was the first underground shaft in a British metalliferous mine to be worked by electricity.

The Greenside electric loco pulling tubs into the Lucy Tongue Level, circa 1920. The man on the extreme right is Captain W H Borlase, and the man next to him J Oglethorpe. Below: man tipping a tub of ore into the hopper outside the Lucy Level. The hopper fed ore directly into the primary crusher.

The power house, built in 1891 to provide the mine with electricity. The pipe brought water down from a leat to turn a pelton wheel which provided motion for the generators. Below: circular buddles, used for separating fine particles of ore from waste material.

Blacksmiths three — J Oglethorpe, J Brown and G Brown outside their smithy, circa 1920.

6 From Ponies to Locos

In 1875 the mine came under the experienced eye of Captain William Henry Borlase, son of a Cornish Mining engineer, who was promoted to the position of manager as a result of the untimely death of Richard Grenfell at the premature age of forty-two. The Borlase family had come over to England from France in 1066 and were one of the oldest mining families in the country.

A great area of the mine both above and below the Lucy Level was now ready for stoping, and there was still plenty of ore yet to be won from near Hick's Sump, below Lucy Level, and Grenfell's Sump, above. The taking of this, however, was not a priority as the price of lead had dramatically fallen to only £9 a ton. Undeterred by the recession, Captain Borlase set about making the best of this quiet time by putting his house in order.

A strong personality with an astute business acumen, he set about laying his recommendations to the board. During this lull it would be advantageous to continue with as much development work as possible, and also some modernisation of the plant in preparation for when the market was once again buoyant. His biggest and most revolutionary idea was to turn virtually all the mine over to hydro-electric power, which would drive winding gear, fans, pumps, and lights. But perhaps the most startling idea of all was to install an underground electric railway system for hauling the ore. The directors were suitably impressed with the plans and, bearing in mind the vast amounts of ore which were lying in reserve, they agreed that, revolutionary as the ideas were, Greenside had never lacked a forward thinking plan. And so Captain Borlase was given the go ahead.

The power for the operation, which would create the necessary 600 volt DC supply, was from the top power house which was built at OD 1,300 feet near the junction of Red Tarn Beck and Glenridding Beck. The power house contained a Gilkes vortex turbine of 100 horsepower driving an Immisch four-pole compound dynamo. This whole system was in fact duplicated for increased capacity for the future, should the need ever arise — such was the careful planning of the operation.

To ensure there was enough water to power the turbines, existing dams were renovated and rebuilt. Red Tarn, a natural tarn under the towering eastern face of Helvellyn, was the site chosen for one of the dams. This was constructed of earth and faced with stone to a height of twenty feet. When full it would cover an area of around 22 to 25 acres, and a spillway was incorporated for safety, with a 15 inch main outlet valve.

The second dam was already situated at Keppel Cove. Here, nestling under the conical shape of Catstycam's northern ramparts, the natural formation of the moraine made it eminently suitable as a dam site. In its construction it differed from Red Tarn in that more natural debris was used. In fact this site had been used earlier by the company when they had diverted Brown Cove Beck into the tarn on the west side and put a spillway in to retain a reasonable depth of water. The dam was now upgraded and the construction improved to increase capacity to around 6.2 acres, the height of the dam creating a water depth of about thirty feet.

Just above Keppel Cove are the remains of the incomplete Brown Cove Dam at OD 2,000 feet. The so-called Top Dam, feeding Swart Beck, did not play any direct part in the hydro-electric scheme, its reservoir being mainly directed to the washing floors and underground to the hydraulic engines. The main water supply for the project, therefor, was taken from the Helvellyn and Catstycam watershed, and an important leat was constructed from Keppel Cove Dam where it met with Red Tarn Beck. The leat was superbly engineered, wood-lined for much of its length, and on steep and exposed sections of the fell it was in fact covered to prevent debris rolling into it, thus ensuring the flow of water. On the final stage of the leat, much was cut through solid rock and here can still to be found the old sluiceman's hut, complete with fireplace carved out of the rock. The wooden reservoir is four-hundred feet directly above the whitewashed power house, and from here the water was released through the sluice-gate into the cast-iron pipes which blasted the water to the turbine.

From the power house, two unsheathed one-inch diameter copper cables were taken overground suspended on wooden poles to the Low Horse Level. Here, in the interests of safety, the cables were lead covered and were attached to the level wall on timber poles. The cables were then hung the 480 feet down Willie Shaft to the Lucy Level. Willie Shaft was now virtually obsolete, although the water pipe from the reservoir in High Horse was of course still being used, and generated pressure to power the hydraulic engine for the winding gear in the Lucy Shaft, which was now down a total depth of 60 fathoms.

Because of the volume of power required for pumping, hauling, drilling and ventilation, a further two 9 horsepower series-wound motors were installed, one to assist at Smith's Shaft to provide air for two rock drills, and another in the Lucy Shaft, a flow pump which pumped water from 60 fathoms below the Lucy Level. These new engines also replaced the ageing Cameron pump which was driven by compressed-air from the Lucy Adit, nearly a mile distant. This had been somewhat inefficient since the introduction of the air drills, as it could not cope with winding, ventilation and drilling as well. This area of the mine was to have the distinction of being the first to have electric shaft-winding gear.

The Smith Shaft itself and the area around was blasted out with the aid of the newly introduced air rock drills. Once started the shaft progressed quickly, but the drills were the earliest type and the steels were solid with no water jet to cool or lubricate them. Consequently, an excessive amount of dust was generated. Noise too was part of the miners' hardship, but none of them had ever experienced the horrific din created by the new drills. It was because of the volume of noise and dust generated that the new type of drill became known as the "widowmaker".

With the advent of the new dynamite explosive with the slow burning fuses, blasting went forward at a greater speed and in a much more uniform and controlled manner. The mine had been using the new explosive since 1887, although during 1891 the company was still purchasing gunpowder as well. Suppliers of all types were delivering goods at this time, again suggesting that many alterations were still being pursued.

For surface work for instance, Joseph Bond Ltd supplied numerous sets of wagon wheels, whilst Joseph Bowerbank supplied sixty furnace bars and four new furnace door frames.

Another first in this century was, without doubt, the fact that Greenside was the first hard rock mine in Britain to have an underground electric loco. The first driver of this unique piece of equipment was Thomas Allison and for this job he was paid £4.16s.0d a month. Other new jobs created were given to William Hurn and Joseph Shaw Jnr, working in the newly built electric plant. It was the responsible job of George Mark Lancaster to turn on the water each morning at the sluice, thereby setting the generators going.

The price of progress even in those days created cuts, and the working days of James Kirkup, the horsekeeper, were numbered. The loco was capable of pulling twelve loaded wagons weighing some eighteen tons. This eventually would replace many of the horses, although not all, for horses continued to work at the mine in a limited capacity for years to come.

The actual date of the loco's commission is somewhat in doubt, although in December 1891 a maintenance report states that no more than £25 had been spent on it, and as yet, the company had not engaged a qualified electrician. Due to the outstanding reliability of the loco, however, six horses were sold.

It is difficult to imagine the complexity of preparation required to electrify the mine and the problems encountered. Between Lucy Shaft and Willie Shaft a fixed transformer was installed to reduce the pressure from 600 volts to 250 volts, and nearly two miles of phosphor-bronze wire for the overhead cable had to be erected, from which the loco would pick up the power by its four contact pulleys. The power of the engine was such that it could pull its full load up gradients of up to 1 in 20, whilst negotiating the numerous bends with ease. This meant that miles of new track would have to be strengthened and very likely much was relaid to fit the gauge of 22 inches. The loco that Thomas Allison drove was the 0.4.0. and was 32 inches wide and

46 inches high. It had three reductions for gearing, the motor was series-wound for a difference of potential of 200 volts, and it developed 15 brake horsepower. It was built by the General Electric Power and Traction Co of Kentish Town, London. During the next forty years or so this loco was to trundle in and out of the Lucy Adit thousands of times, dragging behind its rake of tubs.

Underground 1895

Miners	97	Caging	1	Tipping	2
Brakeman	1	Electric plant	3	Locoman	1
Pumpman	1	Compressorman	1		

Surface

Joiners	6	Blacksmiths	3	Sharpeners	2
Horsekeeper	1	Carter	1	Labourers	9
Sluiceman	1	Shepherd	1	Cleaner	1
Caretaker	1				

Smelter

Smelting	8	Roasting	2	Calcining	2
Labourers	3	Reducing	1	Weighing	1
Separator	7				

The coming of electric power also meant that all the winding stations, shaft tops, turbine house and caging stations were lit by six 100 volt lamps. Further to this major step forward, the turbine house and important underground stations were connected by telephone.

During this period of advancement and progress, however, success was marred by a fatal accident. On Friday the 13th of February 1891, Robert Slee, a banksman, was working with Thomas Pollock, a forty-four year-old wagon cager in the 40 Fathom Level off Lucy Shaft. At about 12:00 noon Robert was working the windlass and Thomas was pushing the left hand trap door back but, as he did so, he slipped and fell the twenty fathoms to the bottom of the shaft, dying instantly of a fractured skull and leaving a widow and three children.

For many years the miners had paid 1/- per week into a welfare club, to help miners' families in need, as in these times it really was a case of "no work, no pay". Strangely, in this instance, a Mr Richardson, acting for the widow, approached the company for compensation. Solicitors Cant & Fairer of Penrith, acting on behalf of the company, offered Mrs Pollock, on the 16th of March 1891, the sum of £50. A letter disclaiming all liability with regards to the accident was required, should the money be accepted, which it was. Although this amount appears small by present day standards, it would represent about a year's wages at that time.

The following year, in June 1892, four men were sinking a shaft when a large iron kibble fell and killed one of the men, Christopher Jeffreys. Tragedies such as these were not common in Greenside Mine — but occasionally fatalities did occur.

The total workforce at the mine in April 1895 numbered 157 (see table above), and the total wage bill for the month was £760.10s.0d. This does not include directors' or managers' wages.

The Lucy Level was still being driven forward and was now in approximately 1,100 fathoms, just a few fathoms past the area of the planned Murray's Sump (not to be confused with the later Murray's shaft). William Hugill and five partners, and also John Bainbridge, were driving the level forward.

Captain Borlase was a resourceful and strong-willed man to say the least, and was not to be trifled with. He was also a man of ingenuity, as the following case illustrates. Some winding gear was required in Smith's Shaft, although why the existing equipment was not suitable is not clear. The message was clear from Mr Borlase, however, as instead of a plan he handed over to four of the miners a simple drawing of a winder which he had sketched himself. Telling them to find some scrap and knock it up, he would find some wheels when they had sorted out what was wanted. The four men, Tom, John, Jos and George set off on their project, intimating that Borlase was the chief comedian at the mine! There happened to be an old winding engine on the old Low Horse Level at the top of Willie Shaft, a grand engine but with no water to drive it (the water had escaped out through the old workings), so from it they used the iron shaft which was over thirteen feet long. It had been forged out square and then drawn under the hammer on the corners and made octagonal. With a 6ft 6ins brake wheel picked up from the Old Low Level incline and two 5 foot winding drums, the job was going well. Off the four went to lower them down the 480 feet to the Lucy Level below. All went well until a loose drum went over the edge and the chain suddenly broke. Away the drum went, smashing and tearing at the old wooden shaft timbers and ripping out the pipes as its momentum grew. Smashed to pieces, the centre of the loose drum was the only part unbroken.

Undeterred, the men carried on with the job. The next thing was to get casting two drum sides to run on the old centre-piece, and two discs with opposite pin holes and octagon centres to fasten on to the octagon shaft. In the meantime, Mr Borlase had got the wheels. Then to fixing, and as cement was not much in use in those days, heavy cross sections of timber were laid down and fixed with bolts. Then there was the motor, a huge affair made by Jackson's of Manchester, 500 volt four-pole with a 2 foot diameter armature. After a good deal of hard graft, jacking and pinching and scheming, it was got together. Sweat and blood, but no tears ("nowt gars reet till somebody's got napped and a bit of blood let owt"). And so the first electric winder was put into commission in this country.

Electricity was still in its infancy and no-one had hit on the idea of a reversible motor, so they came by a double-clutched, bevel-geared reversing device which was coupled to the motor and first reduction gear. This device is still on the property and was in daily use until the old firm closed down.

It was a job well done by real craftsmen. They didn't know who Euclid was and cared less, but practised his art. The shortest distance between two points is — get at it! A centre line, a bit of string, a ruler, and no "top hatter with a rotten plan!" (this winder worked continuously until the 27th of August 1941 when the drum shaft twisted off the gear side, leaving the drums standing intact with one cage up and one down).

Considering the enormous size of the stopes created over so many years, the many hundreds of feet of vertical workings, and the large workforce employed, serious accidents had been minimal. Perhaps it was due in part to the recent mechanisation, the diligence of the miners, or just good fortune. However, this run of luck was to be broken.

On the 9th of April 1901 Joseph and James Watson, W Nanson, and John William Lancaster were sitting down having their bait. They had just settled down when, without warning, the floor beneath them suddenly gave way. Joseph Watson leapt and scrambled for his life as he felt the ground go, but his brother and the others plummeted into the black abyss below. Miraculously, James Watson survived with some internal bruising and W Nanson suffered concussion from falling debris. Alas, fate did not spare John Lancaster who was aged thirty-seven years at his death and was laid to rest in Patterdale churchyard.

A year later occurred a mystery that will never be solved. On the evening of the 1st of December 1902, miner George Chappelhow went up to the mine shop to purchase some items. It was around 8.00 pm and black as pitch outside, when he went to ask the foreman of the washing floor if he should go and tend the sluice gates. He was told not to go. By 10.00 pm the weather had grown worse and the rain was teeming down, yet George donned a coat, lit a lantern and left the warmth of the bothy, informing his mates he was off to check the sluice. The sluice was nearly a mile away up at Keppel Cove, a lonely walk at any time, but on a winter's night, with the wind howling and the rain lashing down Brown Cove, it would have been a most inhospitable and frightening place. When midnight came and George had still not returned, a search party was formed and sent out but it was useless. The weather in the high ground was abominable and the attempt was abandoned until dawn when a full scale search took place. The sluice he had gone to turn off was still open, and not far away his miner's stick was found (many miners carried sticks for good reason — they were used to help keep their balance when they walked the rails in the long horse levels, as there was often two or three inches of water on the floor). The search continued, but there was no sign of George, and it was assumed that he must have slipped into Keppel Cove Tarn, which was covered in ice, in the inky blackness.

The awful truth was discovered when the ice melted and the tarn level dropped and his body was found and recovered from the water twelve days later. George was forty years old, and unmarried. His body was interred in Patterdale churchyard.

After all the activity over the past few years, things started to get back to something like normal. Electricity was now being taken for granted, as was the new compressed-air drilling. Despite the dust created, it was obviously more efficient, and the miners themselves experimented with all sorts of ideas to lay the choking dust. The simplest method was to throw water on the face and the drill, and it did help matters, but it was time consuming and water in the quantities needed was not always available.

Years after the air rock drills were introduced, the practice of blasting with black powder first, to loosen the ground, and then using the more powerful dynamite explosive was carried out. This, the miners discovered, was a more efficient way of breaking very hard rock. This method dislodged a larger quantity of rock and in larger lumps because of the rending by the powder. It produced less galena fines for the mill, thereby resulting in more efficient concentration. It is not clear when this practice stopped, but it was certainly well after 1900.

The smelter continued to work until 1918, although the practice of extracting the silver from the metallic lead ceased in 1903.

It is interesting to note that around 1910, one or two Austrian and Italian miners came to seek work at the mine and by all accounts soon earned the respect of the locals. Some of the names were Pala, Levino, Mignonelli, Vallerri, Gipio and Lenzi. Many were already expert miners and adept with the air drills. Some, like Pala, Mignonelli and Antamata, gave their names to rises and levels, and many of them settled in the village making homes and raising families.

The development work continued in the sole of the Lucy Level, Hick's Sump now being down twenty-six fathoms. Along the Lucy Level, fifty fathoms from Hick's, was a new sump called Shaw's and this had already connected down below with the 40 Fathom Level. Some sixty fathoms on from the collar of Shaw's, a gang was busy sinking Murray's Shaft (not to be confused with Murray's Winding Shaft). Eventually this would connect with the 40 Fathom Level. In the sole of the 40 Fathom Level itself, between Shaw's and Murray's, a new work was being blasted down which was to become known as Edward's Sump and was being sunk by George, Thomas, John and Robert Shaw for £12 a fathom. This eventually would hole out into the 60 Fathom Level which was the bottom horizon of the Lucy Shaft.

The method of working the lode was called "middling", used because of the width of the vein and the lack of very large timber readily available. This method had been used for years and had proved very effective. First of all two winzes, 9 feet by 6 feet were sunk between the two main levels and about 150 feet apart. Six feet above the roof of the lowest level, a sub-level, 6 feet by 6 feet was driven to join the two winzes. Along this level, about every 24 feet, holes were cut for hoppers, through which the ore was dispatched to the mine tubs waiting in the lower level. The miners then commenced overhand stoping from the sub-level, blasting upwards to remove the vein. As the ore fell it was used as a platform to work on, and drawn as required.

The Low Horse Level was virtually redundant and had started to fall into dereliction. Virtually all the ore drawn was now trammed out from the Lucy Level by the electric loco to the mill, which was situated 150 feet below the crushing and dressing plant.

On completion of the milling process, the concentrates were sent to the smelter where 16 cwt batches were roasted in the reverberatory roasting furnace. From there the ore was smelted in Scotch hearths. The lead was cast into pigs of about 112lbs and sold as silver lead, the silver content being about 13 ounces per ton. The slags from the hearths were smelted again in a rectangular blast furnace, and any residues caught in the mile-long flue were retrieved when the flue was entered and swept. This ghastly job was done about twice a year and on collection the residues were returned to the Scotch hearths for smelting.

In 1910 Smith's Shaft reached new depths at the 90 Fathom Level, and a sump was blasted below to collect water. The level off the bottom of the shaft was driven north and south over a considerable distance, and ore was found in both headings and also in the sole. The miners followed the ore body lower, and with every fathom it seemed to be improving. Consequently, it was decided that a new shaft adjacent to the Smith's Shaft should be sunk. This was to the north on the 90 Fathom Level and was to become known as the New Skip Shaft.

The levels driven out from the bottom of Smith's went a hundred fathoms south in productive ground, whilst to the north a similar distance was driven, much of this in good ore both above and below the level. As this again proved the vein, it was decided to sink the Skip Shaft to the deeper position of the 120 Fathom Level, with the collar winding gear in the 75 Fathom Level just six fathoms north of Smith's Shaft. The Skip Shaft was sunk following the footwall of the vein, and by June 1910 it was down twenty-two fathoms, half its intended distance. Unfortunately, this was not a successful venture, the shaft being too small and somewhat ineffective as the ore, once raised to the caging decks at the 75 or 90 Fathom levels, created more work. The heavily laden tubs being handled twice in the small area was time consuming, and a more efficient method of hauling had to be arranged.

It was therefore decided to sink yet another shaft, but this time the shaft top would be on the 90 Fathom Level and sixteen fathoms to the north of the bottom of Smith's Shaft. The proposed new shaft which became known as Murray's Shaft, was driven at an angle of 76 degrees following the underlay of the vein, and of much larger dimensions than the Skip Shaft. The company also intended to install an underground crusher to ensure maximum and uniform loading of the tubs going up to the mill.

Due to the enormity of the project it was to be many years before completion, although the shaft would be used as sections became available. The mine depth at this point was 540 feet to the bottom of Smith's Shaft with another intended 660 feet making a total vertical depth of 1,200 feet below Lucy Level. All fresh air was blasted down in galvanised trunking, but here the mine was dry and the air was heavy with choking dust which would be drawn into the lungs. Years of working in these conditions affected miners with the disease known as phthisis.

Whilst stoping ground south of the Skip Shaft, miner Alexander Cummings, just twenty-four years old, was badly crushed by a large rock. He died from his injuries on March the 29th 1910.

The 1914-18 war approached and, as everywhere, young men in the dale answered the call to arms. Dozens of them left for the fields of France, many never to return. Greenside carried on its lead production which was required for the war effort. Production actually dropped to half that of the preceding five years, and the average output for the war years was a mere 663 tons compared to the previous average of 1,335 tons. This was probably due to the acute labour shortage. In fact in 1915 eighteen tons of lead were bought from Threlkeld Mines Ltd, possibly to make up the war effort quota.

During 1918 some of the old hands still at the mine were W Dover, W Pitchford, R Rogers, W Brown, W Crawford, C Pala and E Parker. All were receiving around 25/- to 30/- per week. It was during this year that the smelter was shut down, never to work again.

By 1919 the mine was struggling, and although ore was available the deplorable state of the world market was bringing the mine to its knees, and as the year progressed matters continued to become more depressed. The directors were left with no alternative but to go into liquidation. Papers were filed on the 17th of November 1920 and an inventory was taken of the company's assets.

Some of the more interesting items were as follows: 48 mine wagons, 577 fathoms of rail at 29 pounds per yard, 525 fathoms of rail unspecified weight, 500 fathoms of rail at 50 pounds per yard, four Holman drills, two Holman hammer drills, one Holman telescope machine, one Ingersoll telescope and one Meco telescope. Waterwheels: one: 20 feet by 3 feet, one 16 feet, two 30 feet, three 6 feet. Four Scotch hearths, one refining furnace and one slag furnace. There were also pumps of every description, cables, air trunking, steel pipes, electric motors, Pelton wheels, buddles, jigs, crushing rollers and turbines.

Somehow work was carried on and production of a sort continued, albeit somewhat erratically. The country was still licking its wounds and recovering from the worst war the world had ever known, and it took until the 6th of February 1923 before a new company was formed with a cash injection of £35,000. It was 1925 before things started to get back into some sort of order.

Many things had changed in the last few years and perhaps the most radical was the closing down of the smelter. All the concentrate was now taken by a traction engine called *Helvellyn* to Troutbeck Station and from there transported by rail to Newcastle for smelting.

The large volume of mill tailings was a problem which would never go away, and the newly appointed manager, Mr Edward Thomas Borlase, (W H Borlase's son), had considered the possibility of using some of the mill waste to make an early type of breeze block. Whether or not this was successful is not recorded, but perhaps this is where the Coal Board got its idea from! In some of the Nenthead mines blocks made from tailings have been used to arch areas of the levels. However, at Greenside it was obviously never pursued as a commercial project.

Mr Borlase had taken up residence in Greenside Lodge with his wife and three children, and his journey to the mine was made on a fine horse. The local children would wait at the gates on the road which they would open as Mr Borlase threw them pennies.

Workforce at 16th of May 1925

Miners	36
Timbermen	4
Labourers underground	7
Labourers surface	11
Washing boys	4
Washing plant	11
Electric plant	5
Cagemen	6
Fitter	1
Mechanic	1
Mason	1
Joiners	2
Blacksmiths	6
Drill sharpener	1
Block makers	3
Horsekeepers	2
Loco driver	1
Cleaner	1
Carters	3
Postboy	1

The foremen or undermanagers at the mine were at this time Hedley Grenfell, George Webster, William Richards, M S Place and G B McKenzie, all of whom were paid between £6.6s.0d and £8 a fortnight. The average working week for the men and bosses alike was forty-eight hours. Of the blacksmiths, John Oglethorpe, Joseph Brown and Harold Oglethorpe were paid £4.16s.0d, while some of the top machine drillers like Antonio

No. 1
The Greenside Mining Company. Limited.
Certificate of Shares.
No 1 to 1200 inclusive.
Issued to Robert Bradshaw Smith
of Penrith, Grocer
Entered on Register of Members.
Thos. Taylor Secretary.
Certificate (Delivered.)
To R.B. Smith
This 14 Day of Feby 1890
R.B.S.

No. [illegible]
The Greenside Mining Company Limited.
Capital,
£76,800 in 7,680 Shares of £10 each.
This is to Certify that Robert Bradshaw Smith of Penrith, Grocer — is the Proprietor of 1200 — Shares Numbered [illegible] to [illegible] inclusive. of £10 each in The Greenside Mining Company, Limited subject to the Memorandum and Articles of Association of the said Company and that there has been paid in respect of each share the sum of £8.
Given under the Common Seal of the said Company the 14th day of Feby One thousand eight hundred & ninety
R.B. Smith
John Pattinson
Directors.
Thomas Taylor Secretary.
N.B. This Certificate must be deposited at the Office of the Company before a Transfer for the whole or any portion thereof can be registered.

Greenside Mining Company share certificate

Mignonelli, Charles Pala and Levino Largioliah earned as much as £13.0s.0d. The new loco driver, Clarry Jenkinson, took away £4.16s.0d. Thomas Hind was the shaft top man at Smith's Shaft, a job he was to keep for many years. Wages were distributed by the company secretary, Mr Mathew Place, a most religious man who, on handing out the wages to the men, warned them about the rigours of drink.

The mine was developed once again, and driving progressed north in the 120 Fathom Level, in which Webster's Sump was being sunk by Seymour Webster and two partners at £21.0s.0d per fathom. The 135 Fathom Level was worked south, and also large reserves of ore were stoped out to the value of 30 cwt a fathom. Just north of Murray's Shaft on the 150 Fathom Level similar values were being won. The mine now appeared to be getting back onto an even keel.

To help with the development programme, four new drills were bought by the company on the 30th of July 1926. These were 4N70 Ingersoll Rand Drifters which cost £74 each and were to replace some of the ageing piston drills. These drills were powered by just 70 pounds of air pressure which, while sufficient, was not as good as it might have been. The development of the mine continued and the lowest working, off Murray's Shaft, was the 135 Fathom Level. The water generated at the mine was around fifty gallons a minute, which is a very small amount in comparison with other Lake District ventures.

It was about this time that the company began paying wages on a fortnightly basis, the money being collected from the Midland Bank in Penrith where its departure was supervised by the manager Mr Harrison. The amount was transferred in cash, an example being the entry for April 1927 when £629.7s.5d was collected. The mine manager with one other person used to travel to the bank by car, and it has been said that he carried an old service revolver for protection. This is probably quite likely as Mr Borlase had seen active service and had reached the rank of captain in the army. There was no Securicor in those days!

7 Disaster at Keppel Cove

On one particular Friday, the 4th of November 1927, up at Keppel Cove dam the water level was noticed by the sluiceman to be low, so the sluice gate was closed to bring the water up to a working level. That evening the weather pattern altered dramatically and a sudden and vicious cloudburst of intense ferocity was unleashed on the whole of the Lake District. The nearby valley of Kentmere was flooded to many feet and a massive landslide at Hay Fell prevented the LMS running trains between England and Scotland. The road at Oxenholme, near Kendal was torn up whilst in Kendal itself and at Endmoor many places were flooded.

High up at Keppel Cove, enclosed as it was by the surrounding mountains, the ferocity of the winds turned the reservoir into a raging, tumultuous force and the waves whipped up by the storm crashed over the soil banks of the dam. The volume of water was such that large erosion channels were soon gouged into the dam. It did not take long for a breach to occur.

It was about 2.00 am on Saturday the 5th of November when thousands of gallons of water spewed into the gill, surging downward into the valley towards the unsuspecting, sleeping villagers of Glenridding.

The first object to get in the way was Rattlebeck Bridge, which was destroyed within seconds, the tidal wave by now carrying with it uprooted trees and boulders. Eagle Farm was flooded to a depth of five feet, a tea hut vanished and debris was even found at Side Farm across Ullswater. At Rattlebeck Cottage Mr Parker, his wife and two children, and their lodger William Armstrong, had been woken by a tremendous crash as a tree was hurled through the house wall, allowing the water to pour in. Struggling in the pitch dark, the wind howling and rain lashing down, somehow Mr Parker and William were able to save the family.

The Millcrest Hotel (now Glenridding Hotel) was extensively flooded, and the force of the water was so intense that four girls who had been asleep in a lower room floated to the ceiling on their mattresses. Their screams for help brought bootboy Ernest Thompson to the rescue, just in time to save one of the girls from being swept through the window. He managed to wake Mrs Ridding, the assistant clerk at the Post Office, to telephone and raise the alarm.

The amount of damage could not be seen until morning when it was found that the main road bridge at Glenridding, which had taken the full force of the torrent, had withstood the battering. Miraculously there was no loss of human life, but the valley sheep and cattle had fared badly, dozens being lost. Enormous damage had been wrought, and as the day wore on the full extent was revealed. In just a few hours ancient dry stone walls were down, bridges and houses severely damaged, and the main street and shops were filled with mud and debris. Dead animals were strewn about along with toys, pots, pans, carpets, clothing and other domestic accoutrements. Nothing had been spared the devastation. The Millcrest, famous for its antiques, lost its most treasured pieces. A 150-year-old grandfather clock and an oak cabinet dated 1707, an heirloom of the old Mounsey family, were smashed to pieces. Walton's store was totally wrecked as was the Co-op store.

Meanwhile, totally unaware of the carnage below, George Birkett struggled through the driving wind and rain at 7 o'clock as usual, to tend the sluices and turn on the power. To his horror, on arriving at the dam he was confronted with little more than a very small pond and the dam laid totally waste, breached by a gap thirty feet wide and forty feet deep.

The alarm raised, a team of thirty men were dispatched from the mine by Mr Borlase to start the clean up in the village. Soon dozens of people were involved as the small community got together to tackle the work.

To help the victims of the disaster, a fund was set up and in the first week £24.13s.7d raised. Penrith Urban District Council offered help and many personal pledges were given. Mr Roberts, the chairman of the committee on behalf of the victims, started to ask for reasons why this disaster should have occurred.

The mine had now lost its main power supply and the supply leats were in many cases washed away, making the operating of the mine virtually impossible. This fact was soon realised by the locals, who were well aware of the added hardship which would arise should the mine close. Greenside employed almost everyone in the village either directly or indirectly, and had long been an integral part of village life.

Responsibility for the disaster was laid squarely at the offices of the mining company, damage was assessed, and the claims came rolling in. The main road bridge, although it had remained standing, was now considered unsafe after being surveyed by Mr Holmes of the Highways Department and he recommended that it be rebuilt and widened. All sorts of claims were made against the company and total losses were estimated at over £25,000. The company paid all the claims but this, together with the falling lead price, put such a strain on the capital resources that it went out of business.

In order that the mine could function, it was essential that the dam be reinstated with all possible speed. The site was inspected and it was decided not to repair the old dam but to build a concrete replacement instead. This would be erected across the Brown Cove valley, just below the site of the ill-fated Keppel Cove Dam. The new construction would be 42 feet high, 6ft 5ins wide at the top, 258 feet long with a 17 foot wide spillway. The actual base of the dam was 20 feet wide and this was supported on the external face with a further 20 foot wide buttress, 8 feet high. Into this were set the sluice gates.

The necessary aggregates for the construction of the dam, plus much of the other basic stone, was taken from the northern screes of Catstycam. A small railway was built to collect the shale which was then put into a hopper and mixed with concrete. From here it was transferred down a short incline where the crude mixture was poured into the shuttering of the new dam. The concrete was reinforced with a variety of disused drill steels from the mine. The large spillway built on the south side ensured the dam would not be put under dangerous stress. The sluice gates bearing the manufacturers name, Davison & Spencer of Hexham and dated 1869, are still in situ and were possibly rescued from the original dam. These are, however, not the only names to survive, for two erstwhile miners, J Dawes and W J Richards have left their names in the concrete facing for posterity.

During this period it was decided to upgrade the electric power supply. This was done by first constructing a small concrete feeder dam, just below the miners' bothy. From this ran a mile-long leat which contoured the south side of Glenridding Beck, and just above Rattlebeck the water was sent down to a new power house. Here the water, having a head of 360 feet, fed the turbine which in turn developed 300 horsepower. The electricity generated was taken up to the mine on overhead cables.

Within just over a year the Keppel Cove dam was well under construction under the supervision of Bill Richards. This was dangerous work, for the weather in this lonely valley can change in minutes, raging storms building up while the vale of Ullswater can be tranquillity itself. Violent, gusting winds were the worst enemy, the blasts coming without warning — there had already been one fatal accident.

The total number of accidents reported during the work so far totalled eight, the most serious of these occurring on in November 1928. This involved the driver and his mate of the big Fordson tractor. The wheels of this monster weighed nearly a ton, and it was used for all the heavy work, bringing up the timber and bags of cement. The two men were lucky to escape when the lumbering vehicle plunged off the track and fell on its side. John B Graham suffered concussion while James H Storey sustained a leg injury. Apparently unshaken, James was back at work in less than a week when once again the Fordson slewed off the track and turned over. But this time James was killed outright.

Work pressed on and by April 1929 there had been a further eight serious mishaps including two fatalities. This had certainly been a miserable place to work and the men were glad when the job was done.

In accordance with a recent Act of Parliament, the company's three main dams were inspected by a Mr E W Dixon and his assistant from Leeds, for which they were paid £10.10s.0d and £1.1s.0d per day respectively. This did not include any expenses, and they were the most reasonable surveyors available. The new Keppel Cove dam received a clean bill of health, and the other two passed, but certain modifications were recommended. Red Tarn was to be reduced in depth by 2ft 6ins and the Top Dam by three feet. A total of £500 was spent on upgrading them. Further to this, the company instituted a private investigation into the cause of the breach, and this was carried out by Herbert Lapworth and a Professor Poole. They stated in their findings that it was considered the actual dam was generally of sound construction, and although moraine had been used as the base, this had been faced with stone. No-one could have foreseen the severe weather pattern which had devastated the area on that fateful night. The estimated rainfall at the height of the storm was 3.38 inches in one hour, with swirling winds gusting at 90 mph. This deadly combination had literally heaved water at the dam, eroding the facing and, in a very short time, with the increasing pressure on the failing structure the dam was breached.

Around this time an old steam traction engine, the *Sentinel*, was being used to take the lead to Troutbeck and bring back coal and coke. It had been used for many a year together with its sister machine, the *Halley*. Both were big and cumbersome things, and responsible for the demolition of many a dry stone wall in their day. The end came for the *Sentinel* around 1931 when the driver Archie Graham and mate Johnny Pool were on board, coming from the Keppel Cove direction down the steep track by the gill. Just before the mine offices, the vehicle left the track and the two men had to jump for their lives as the old engine crashed the sixty feet into the gill, its boiler hissing and spitting as it sank into the water. It lay there for nearly sixty years until an enterprising gentleman hauled it out and restored it to full working order.

It seems that somehow, someone got their sums wrong, for during 1931, the unthinkable happened when the new Keppel Cove Dam breached on the north side. The bedrock on which the dam was sited had been either very shallow or non-existent. But in spite of the rupture, which was some thirty feet wide in the dam base, ninety-five per cent of the dam miraculously held. Fortunately, the deluge being partially restrained prevented a repeat of the havoc and destruction of 1927. For the village to have had to endure such devastation again would have been unbearable.

Another major disaster occurred in the new Murray's Shaft which now connected the 90 Fathom and 150 Fathom levels, and was the main haulage and caging shaft. It had, without warning, crushed in from the collar on the north side for the top fifteen fathoms. The steel girders buckled and twisted as tons of back-filling spewed into the shaft, stripping out all the centre timbering as it careered downwards and filling the bottom ten fathoms and the sump directly beneath. Unable to pump, the 150 Fathom Level flooded to the roof.

Eventually, after much hard work and ingenuity, pump lines were rigged through a series of rises north of Murray's. This recent catalogue of disasters greatly contributed to the difficulty of running of the mine. Now, with the water reserve for the electric power gone, the dressing supply now halved and a declining market, for the first time in over a hundred years the mine ground to a halt.

Following the collapse of the company Anthony Wilson, a well known mining engineer from Thornthwaite Grange, Keswick, was, during May 1935, requested to do an evaluation of Greenside Mine and its assets, as it was again in a state of liquidation. His criticism of the old company policy was mild, and being an experienced engineer he realised that the demise of the company was not as a result of mismanagement, but was due to many unforeseen problems. Firstly, there was the compensation paid for the dam disaster, then the second dam failure resulting in a virtual stoppage of production; there was the collapse of Murray's Shaft and the resulting flooding; and lastly, and possibly the most significant factor, the depressed lead market.

During 1914 lead had been standing at £11.10s.0d per ton, but by 1934 it had dropped to a meagre £5.14s.6d. The mine required a stable price of at least £10 per ton to be profitable. The lead price had been poor for so many years that little had been extracted, but there had been enough stockpiled for the mine to just keep going.

Despite all the problems Greenside had always been a good mine, and what was required was a major cash injection of around £20,000. Further development of the mine to exploit the lower ore reserves was essential, and to do this Murray's Shaft needed to be taken down to 165 fathoms. This would give estimated reserves of 72,000 tons — at least thirteen years work.

By modern standards the existing mill was now old and labour-intensive. It was considered too expensive to repair, and a new mill would be a sound investment for the future. This would mean installing a modern plant which would increase the recovery of lead from 10% to 12%.

It was recommended that the 90 Fathom Level should be driven a further 130 fathoms, as lead had been taken from the area above by the old men. The 120 Fathom, 135 Fathom and 150 Fathom would be extended north if the potential of the vein was seen in the 90 Fathom.

Not all the plant was redundant, and some of the electrical winding gear and pumps were in excellent order. Other assets of the company were twenty-seven cottages leasehold, twenty-four cottages freehold, three houses and five acres of land valued at around £8,500. A conservative estimate of this valuation today would be something like £4,000,000.

The starting capital required by the new company was £20,000, but this would not be required all at once for, as mentioned before, the mine was not totally out of commission and there were proven ore reserves. Things were, should we say, just a little rusty.

At the "Jerry" (now the Traveller's Rest), the local ale house, things were anything but rusty. Many a hard working miner had quenched his thirst here. And although it was an ale house only, and without a spirit licence, the occasional nip was to be had. To stop this illegal practice, two plain clothes officers were dispatched from Penrith to investigate and nip the practice in the bud. The night they chose for their visit was the evening of a fox hunt of which the locals were most supportive. The officers mingled and were made welcome on this night of frivolity, and as the evening wore on, the investigating force succumbed to the hospitality and returned to Penrith the following day with nothing to report.

8 *The Basinghall Co 1936*

Early in 1936 the Basinghall Mining Company took over Greenside Mine. Apart from a small collapse at a position 105 fathoms down Murray's Shaft, the mine had stood well. Within a few weeks this area was cleared of debris and totally secured.

An engineer's report of October 1936 indicated that prospects were good, and consequently funds had been raised to ensure the new company could operate. A total of £9,500 had been allocated for sinking and drifting and general exploration work, whilst other monies were set aside for new equipment and reserve. It is also mentioned in this report that money was put aside to the total of £10,000 for a new adit. The balance left was around £2,500. After some initial exploration work it was confirmed the vein could now yield as much as 14,000 tons giving a return of £105,000 per annum. Should the ore body continue in strength 500 feet below the present workings, it would be possible to float a public company with a capital of £250,000 giving a flotation profit to the syndicate of £105,000.

Things started to move fast, and the whole mine from top to bottom was to be refurbished. Part of the shopping list in 1937 was as follows:

8 cwt of carbide	24 Wolf carbide lamps	24 Premier lamps
60 cups & saucers	60 plates	60 soup plates
60 pudding plates	60 cheese plates	12 milk jugs
7 Nupac mattings	200 coat hooks	1 chipper
3 potato peelers	2 steamers	1 frying pan
2 8" teapots	3 enamel bowls	4 bread tins
12 armchairs	40 wooden chairs	1 electric heater
2,000 Woodbines	1,500 Players	1,500 Capstan
300 Capstan Full	1 can opener	1 set of scales
40 beds 6' x 3'	120 sheets 63" x 99"	120 pillow cases
12 miner's helmets	12 square shovels	12 round shovels
6 3lb hammers	4 4lb hammers	72 fuse lighters
750 Polar dynamite	500 capped fuses	28 pkts soap
3 side-tip wagons	1 pair of wellingtons	

The list goes on: paper clips, typing paper, oil, coke, coal, Barbour oilskin coats, sou'westers, over-trousers, saws, screwdrivers, cement, rails, sleepers, hay, wood, nails, and glue. No doubt the cigarettes were for the mine shop which was set up at Bell Cottage and run by Rubin Robinson, whose wife Marjorie worked in the mine office. It is said that a bell used to be rung from here to let the surface workers know when it was "bait" time and finishing time, and this could indeed be how the house got its name. During this period of modernisation it was obviously appreciated that there were dark days ahead, for the situation in Europe was unstable to say the least and the home mineral market would become essential to the war effort.

The new plant was installed by Huntington and Herbalene Ltd of Glasgow, much to the consternation of an American company who'd hoped to win the order. During this period the original electric loco was still running, although this was soon to be replaced by a loco from Wingrove Rodgers Ltd of Liverpool (British Electric Vehicles). The new loco, engine No. 1192, was delivered during February 1938, and less than two months later a second loco, No. 1215 was delivered on the 30th of April 1938.

At about this time the village of Glenridding was connected to the National Grid and this was good news for all. It now meant that the mine was not totally reliant on its own power source at Rattle Beck. This welcome additional power supply, whilst alleviating some problems, created one of the trickiest: how to feed the new power supply underground. The method was to lay sections of 3 inch diameter, three-core copper wire, armoured cable capable of taking a load of 2,200 volts. This was delivered in five-ton rolls.

To lay this in the open is one thing, but in the confines of the twisting Lucy Level it was a major problem. It was first thought that the loco would drag it in, but this soon proved impossible because of the bends in the level, the cable being so inflexible that it threatened to derail the loco. Every available source of manpower was dragooned into hauling the wretched stuff, including office staff! The operation was supervised by Mr Conners who controlled the movement of the loco and cable by signal blasts on a whistle. Due to the shape of the tunnel, the whistle blasts were distorted and also they echoed, which naturally added to the confusion.

Eventually the cable arrived at the newly constructed underground compressor house near Smith's Shaft. Here the new Joy Sullivan two-stage compressor from America had been installed, its number, 25017, emblazoned in orange on a brass plate. This replaced an old Willie Walker compressor which had been installed many years ago by the old company, and had given much trouble in latter years. The steel cogs had become worn and some kept breaking, making the whole unit unreliable. The new compressor was installed by Thornton's of Wrexham under the watchful eye of their head foreman Mr George Bursley. This would supply power for air conditioning, air rock drills and hoists. The power generated from the 2,200 volt high tension motor provided 175 horsepower, gave 100 pounds pressure and produced 11,000 cubic feet per minute. The new plant would, in years to come, be the pride and joy of Tom Hind ("Scan" to his mates) and in fact completed forty-five years service. Nearby, an old Aley McLeand of Glasgow compressor, which had been installed around 1924, was retained as a standby.

The Wingrove Rogers Company also supplied some smaller 1 ton battery locos for use in the lower levels. They were capable of hauling six tubs and were first installed in the 150 Fathom Level whilst the 90 Fathom and 120 Fathom levels were still worked with ponies. However, with the speed of development, it was decided that the ponies, which could only pull two or three loaded tubs, should be replaced as soon as possible by locos. By now, Murray's Shaft was down to the 175 Fathom Level, which was being driven forward with all possible speed.

For quicker analysis of the ore being mined the company set up its own laboratory and assay office. One of the first sets of figures produced informed the management that from eleven main working areas in the mine, the ore reserves totalled 300,500 tons. With the new method of recovery obtained by flotation as against the old gravity methods of concentration, the recovery would be not less than 12% lead which would yield 45,000 tons of concentrates.

By August 1938 the mine was geared to producing three-hundred tons per day, achieved with a workforce of 175 men, about 110 of whom worked underground. The machine drillers, of which there were about forty, were paid 13/- per day, sixty-five labourers were on 10/- per day, the total daily bill being estimated at £200.10s.0d. Other major expenses such as stoping, milling, shops, and royalties amounted to £310 per day, a total daily cost of approximately £511. With the price of lead then at around £14 per ton, the balance sheet would indeed start to look better. Based on these early figures the Basinghall Mining Company made further plans to increase the output to 500 tons per day, and with the current price of lead the future seemed assured.

By now most of the equipment was new and was not considered expendable. The foremen kept a sharp eye out for any misuse or carelessness with tools. One day, Arnold Lewis and his mate were back-filling an old sump with waste. Time and again they pushed the tub to the hole and tipped it until, by accident, they tipped it a little too far and the tub-top disappeared into the hole. The foreman was summoned and after three hours of hard graft the tub-top was retrieved with not a little bad language and stern admonitions to be more careful in the future. The next load was brought and, inexplicably, this time the whole tub fell into the hole! Needless to say, the foreman was not notified of this. Another tub was obtained and Arnold and his mate blithely continued to fill the hole with waste.

Timber was needed in vast quantities in the mine and up to now pitch pine had been most widely used, but this was becoming expensive. The nearest and cheapest substitute was spruce, which was brought over from Charlton's of Hexham. In the climbing manways and hauling areas Canadian redwood was used. In the old days the sliders in the plumb shafts such as Smith's were of purple heart wood which was as hard as iron. In later years the sliders were made of steel, as in Murray's Shaft.

Over the six month period to April 1939 the total ore produced was 22,420 tons, yielding 12.07% lead. The actual tonnage milled for this period was 23,150 tons, the reason for the difference being that some ore was stockpiled for the reserves, from which some had been removed.

Up till the outbreak of war in September 1939, the mine had worked a five-and-a-half day week operating two shifts. This altered almost immediately to three full shifts: 8 to 4, 4 to 12 and 12 to 8, the mine working flat out for the war effort. The general manager was a man called Mr Purvis, who kept himself to himself rather. One of the underground foremen was Bill Hebson, who had come from Hartsop Hall. The electrical engineer foreman was a Londoner, Dennis Flynn, who by all accounts was a workaholic and expected everyone on his staff to be of a similar type. He was not, however, quite the match for electrician Walter Burnett, a local man. Walter and his mate were working a particular Saturday from early morning and the time had gone well past their one o'clock bait time. They were satisfied with the morning's work and were just off to eat when the voice of authority boomed out demanding to know where they were off to. Without turning round, Walter raised two fingers in the age-old gesture, leaving Flynn screaming about seeing the general manager.

The war progressed and an influx of Italian POWs, some self-patriated Poles, and some French arrived. The POWs were kept in what is now the YHA hut until a proper compound could be built, and the Poles were billeted at Pooley Bridge. The mine took on a fairly cosmopolitan air, and of course, with the POWs came the Geneva Convention and its rules and regulations. One simple rule was that the prisoners had to have a hot lunch. The problem with this was that having got them ready for work and transported them through the system to wherever they were working, after an hour it was time to escort them back for lunch. Walter was entrusted with two such charges, neither of whom could speak English. It was 10.30 one morning and the three of them were pushing a tub with electrical gear along the 175 Fathom Level near the 940 Winze (a recently sunk shaft), where Walter had some maintenance to do. Suddenly, one of the Italians fell full length into the drainage ditch, and as he raised himself out he declared he was dying of pneumonia and demanded to be taken to the surface. By the time Walter had managed to get them to surface, and got the one attended to and the other to his lunch and changed, the day was nearly over. Walter mused on whether or not this might have been a bit of sabotage. Some of the Italians worked well, in particular a man called Fontanna, who claimed to be a stone mason who had been commissioned by Italian nobility. It was he who helped build the big wall in front of the mine office, a legacy still to be seen.

The mine was now being driven on in a new stope, the Alma, which is an extension on the 48 Fathom Level above the Lucy Level. This was the only area to be worked above the Lucy Level. Other main areas of working were the sinking of the 940 Winze from the 120

Fathom Level which was now down to the 200 Fathom Level. In the 175 Fathom Level a new shaft was being commenced which was to become known as the North Shaft. In this area the vein was very strong and eventually revealed a solid run of galena four feet wide. The area became known as the Eldorado and was discovered by the two top miners, A Mignonelli and Bill Brown. It was said these two men could smell lead. Lucy Level was driven on further north, and the 90 Fathom Level was a hive of activity with no less than fifteen separate workings. Exploration was also being pursued in the 120 Fathom, 135 Fathom, 150 Fathom, 175 Fathom and 200 Fathom levels.

More Italian prisoners arrived and the POW compound was now completed behind the hut they had been using. The general workforce was also being increased and this meant ferrying men in from farther afield. An old Bedford truck driven by a man called Bryncombe used to bring men from Penrith. Later, two Austins were used and a bus for the Keswick shifts.

As the production at the mine increased, so did the slimes from the mill. The slimes and waste were a problem, there being little area to dispose of the mill slurry. It was decided to pump them on top of the old heaps and to do this an old Mather and Platt Gwynn pump of 100 horsepower was used. The waste was pumped away from the mill through two 4 inch pipes and then dumped on top of the old Lucy Level waste heaps. Engineer J D Wilson thought the fines would fall into the existing tip and solidify, and in order to ensure that the drainage was satisfactory, a series of 18 inch pipes was inset to allow any surface water to run off safely. However, one pipe was set slightly off and instead of draining, the water ran straight down the tip itself with disastrous results. Within hours, after a heavy storm, masses of grey sludge were on the move, all the time assisted by the gallons of water coming from the drainage pipe. The sludge continued down the tip and commenced covering the stables, the ponies in residence being saved just in the nick of time. Nearby, the mine manager's car was also nearly engulfed. From that day the incident was known as "the day the desert moved".

Mr Bloomfield, the manager, had recently joined the company from South Africa and was soon nicknamed "Old" Bloomfield, although he was a young man.

During this period there was another fatality at the mine. One of the regular jobs was for a miner to collect blunt drill steels from the working areas. The steels were thrown down from the working platform in the stope to the level below, after a prearranged signal had been given. Miner Tommy Howe either misunderstood or failed to hear the given signal and was in the way of a falling steel which struck him on the head and killed him.

Output increased and consequently so did the "desert", and because the mine was now on twenty-four hour working and the waste was being brought out at all hours, the whole area was now floodlit.

It was in 1943 that a team of Canadian Engineers arrived to assist in the exploration of the mine. These men, whilst Army personnel, were also experienced hard rock miners and were probably the same group who had been working at Carrock tungsten mine. They did not stay long at Greenside, but they did help finish and install the steel framework of what was to become known as the North Shaft which goes down to the 217 Fathom Level, though much of the work in this shaft was done by a group of Welsh miners. Murray's Shaft was by now down 200 fathoms, its final depth. Both Smith's and Murray's shafts had sumps beneath them, which should have been regularly pumped clear of water. Now and again, however, they would fill up soon after they had been cleared, and this resulted in some spectacular short circuits as the cage electric gear touched the sump water. As this happened on several occasions, it was decided that the sumps would be deepened. This made at least two men very happy, J Hamilton the mine foreman and Walter Burnett the electrician, who had had to do the resulting repair work every time.

The increased development in the mine was cause for a further loco being brought in, engine No. 2489 delivered on the 15th of May 1943.

The surveying at the mine was constant and it was common practice to accept that the previous surveyor's work was accurate, once it had been drawn up on the office plan. This was not always the case however, as an error found by one surveyor, a Mr Daniels proved. According to the plan a newly-proposed inclined shaft from the 200 Fathom Level would have come out 8 feet to the west of its intended position, and although not an insurmountable problem, a lot of extra work would have been created. From the tone of the correspondence regarding this the management were none too pleased, but fortunately the work had not commenced when the error was discovered.

When the war came to its end in 1945 the men started to return home, and some of the Italian POWs stayed on. Joe Hope returned as a pipe fitter and later became a timberman, following in his father's footsteps. Joe was only 5ft 3ins tall and soon became known as "Tot" to his mates. Whilst working the stopes one day he was bending forward when his prize 5/- Ingersoll pocket watch fell eighty feet into the darkness. Joe decided he was definitely not going down to look for it, but another miner, Isaac Coward did, and found it, still going and with only a cracked face! They wrote to Ingersoll to tell them of the incident but heard nothing back.

It was the policy of the Ministry of Labour at that time to find work for displaced persons, and many of these were real characters. There was Charlie, a crazy Cornishman who worked as a pumpman and never wore shoes or socks, his eccentricity being put down to the fact he was ex-merchant navy. Another such character was drill-sharpener John Kemp who used to chew black twist and leave it on the furnace to dry out. Said it killed the germs and that it would be OK tomorrow!

By now the miners were working in the 217 Fathom Level, and all the drilling was by air rock drills with tungsten tips. In this area of the mine the heat was apparently intense and the mine was very dry, adding to the hardship. These headings were taken forward by driving twenty-two holes across a 7 foot by 7 foot face, and after the charge of Noble's dynamite had been laid and discharged, the heading would normally clear 4ft 6ins. Much of the mucking out in these lower levels was done by hand and not by Eimco loading shovel.

Thankfully, accidents of a serious nature were, as mentioned previously, a rare occurrence, but there were many close calls like the one in 1948 when Jimmy Kirkup somehow managed to fall out of Murray's Shaft cage. Jimmy was blessed with rather protruding ears and as he fell into the sump many feet below, his mates insisted that his ears had kept him airborne. Others claimed, more realistically, that it was the flotsam and timber bobbing about in the sump which helped to break his fall. He was retrieved with utmost speed and survived to tell the tale.

Murray's Shaft, in its time, had been the scene of various misfortunes, one in particular occurring as the cage was descending, taking an eight-man shift down to work. The usual banter and crack carried on as the cage descended with just the glow from the odd carbide lamp. The men suddenly realised, as they approached the bottom station at the 200 Fathom Level, that the cage was not slowing down. There were shouted oaths as the lights of the landing area flashed by, and the men were all suddenly waist deep in the water of the pumping sump. How and why this happened is not clear, but the feelings of the men were certainly made clear to the hoistman of the day when they finally returned, dripping wet, to the shaft top.

On a day when a particularly violent storm was raging topside, two men, Henry Richards and Joe Hope, were carrying some air trunking along the 120 Fathom Level. Joe was at the front with Henry bringing up the rear when suddenly there was a blinding flash and both men were on the floor. Somehow, lightning had come through the system and given them a belt. Henry insisted that it was their rubber boots which saved them both. Joe had another lucky escape when he was going to do some timbering underground. On his way to the job he started to cross a boxhole with planks over it but, unknown to him, the supporting timber had been removed and he crashed through the stope onto the level below. Although badly bruised and dazed he was none the worse for wear.

When miner Bobby Jenkinson was climbing down into a stope to do some work the section of rail which the ladder was hanging from became dislodged from the rock and the rail, ladder and Bobby crashed the thirty feet to the stope floor, narrowly missing Edwin Turner who was working below. Bobby sustained a broken leg and was extremely lucky that the rail had missed both him and Edwin.

In some of the deep levels off Smith's Shaft ponies were still being used under the care of Jack Park, the horseman. Their names were Duke, Star, Kipper and Eric and they were mostly docile and good natured and seemed happy in their dark world. However, a group of men were having their bait one day near the shaft top when there was the clattering and banging of a horse and tub approaching fast. In the gloom they saw an untended horse coming rapidly towards them. Luckily, the shaft barriers were up and the horse was prevented from disappearing into oblivion. The abrupt braking system did not seem to bother it at all, its main concern being for the apple one of the miners was eating!

The ponies were brought out of the mine at night to graze and on working mornings the men had the devil's own job to catch them, and had to bribe them with oats and even chocolate. On Sundays, however, their day of rest, they would come trotting to the fence for a nuzzle. The miners trusted them underground and probably with good reason. On one occasion Arnold Lewis and his mate were bringing some tubs out of the north end of the mine when the horse's ears begin to twitch. Suddenly it bolted, tubs and all, off down the level followed by the two men, and seconds later a large section of the roof collapsed. The animal's sixth sense had certainly saved them from an untimely exit. One particular horse, Star, unfortunately came to the end of her earthly life whilst working underground and was quietly laid to rest down the nearest sump.

By 1950, miners were still wearing cloth caps and clogs with cut-off wellingtons over them, acting as waterproof gaiters for the wet areas. Helmets were available but not compulsory and the men seemed to be content with the traditional gear as they made their way around the mine, shuffling along the railway lines with their sticks for balance. One man who believed that the cap was better than the helmet was engineer Harold Oglethorpe, whose father and his father before him had rived the lead from the Greenside veins. On a routine inspection, Harold was in the Lucy Level and made a turn down a side drive. Not seeing the air compressor pipe his cloth-covered scalp made sharp and painful contact with it and from then on, as an example to the men, Harold wore a hard hat (I can testify to the fact that to walk directly into a four-inch air compressor pipe at normal walking pace, even wearing a helmet, brings one to a very abrupt halt).

One of the biggest problems in the mine was the shifting of the broken ore from the hoppers. The miners stoping above would send the stuff crashing down to the hoppers below, and the large and irregular vein stuff would lodge and sometimes block the hoppers. Attempts would be made to free blockages with drill steels, scrapers and a variety of hand held tools. When all ingenuity failed, a charge of dynamite on a long pole would be inserted between the offending rocks. However, if the pole was not long enough, this meant someone climbing into the hopper, which was of course a very dangerous practise. Tons of unstable material above were liable to move at the slightest jolt, and the exit space was limited. Isaac Coward, a staff man, was in fact seriously injured clearing such a blockage.

Not all incidents occurred underground, and not all were related to work. Electrician Walter Burnett was a real Geoff Duke of his day and reckoned he could beat anybody on his Velocette OHC motorbike, the fastest two wheels in Glenridding. The record, which Walter held, was three minutes from the village to the mine. One morning he screamed up from the village, round the "S" bend and was just getting nicely wound up past the Rakes and onto the powder house. To his horror, right in the way was the morning shift bus coming from the mine. Walter wasn't sure whether the bus was late or he was early as the bike broad-sided and the bus, driven by a white-faced Alan Boustead, skidded. Walter, legs gripping the machine, just managed to squeeze through between the bus and the fast-looming dry stone wall and, continuing at speed past the powder house and Bell Cottage, arrived safely at his parking spot, none the worse for his ordeal. Bus driver Alan Boustead, however, had a few chosen words to say.

In the very early '50s the Greenside company applied to work the old Hartsop Lead Mine, as they were looking for more lead reserves. The workings at Greenside were becoming expensive to work, the mother lode was becoming more elusive and the vein structure was not yielding as much as it had. The planning authority, whilst sympathetic, would not allow a plant to be built, but would give permission for the extra heavy traffic if they wanted to work it and transport the ore by road to the Greenside plant. The costs, even in these early days, were prohibitive.

By now the shifts had returned to normal and the working week was forty-three hours. The men were paid fortnightly and could also earn production bonuses. The workforce was now around ninety men, forty-seven of them underground, fifteen in the mill, and the rest on surface. Fewer men now were staying in the mine hostel as most of the workers were married and settled in the village. Tommy Simpson was the hostel keeper and some of the last men to leave were Danny Wilson, Harry Bruce, Jock Young, Tony Mignonelli and Jimmy Ladyman.

Now that the war was over the mines inspector started to make more routine checks and his requirements became more demanding. The visits became something in the nature of a game of cat and mouse, and as soon as the inspector's car was spotted the mine office was alerted and a series of signals tapped out on pipes warned the men underground of the impending visit. On a check early in 1950 he was certainly in no mood to be fobbed off with tea or the manager's guided tour. The mines inspector informed Mr Bloomfield that the cables supplying power for the loco were a hazard, and insisted that they be completely boarded in. Mr Bloomfield politely explained that this was totally unnecessary as there had been no accidents and the men always walked on the opposite side of the level. The inspector left, his instructions clear, saying that he would return to check that the work had been done. Six months later he returned, unannounced, and of course nothing had been done about the boarding of the cables and he immediately condemned the operation.

Chaos reigned. The old 1938 Lucy loco (No 1192), had to be converted without delay to battery motive power. As luck would have it, a charging unit was located at Halkyn Mine in North Wales and Mr Bloomfield and Walter Burnett went straight away to examine it and arrange its purchase. The same week a visit to RAF MU 14, Carlisle, produced a mercury arc rectifier, 440 AC to 110 DC, and this was also purchased. Walter, along with other fitters, soon had the loco converted and running. Wingrove & Rogers supplied another loco, No.4472, on the 18th of October 1950. It was essential to have a back up loco as the working battery life was only around four hours. The locos used five gallons of distilled water per day, the reserves of which were regularly checked by HM Customs.

In an effort to find more ore reserves an extensive search was made on the northern slopes of Hartside by trenching and the more sophisticated method of a series of diamond drilling test bores. The main rig was built by a Workington engineering company, the drillers were Billie Sinkinson, Ernie Jamieson, Abe Pattinson and W Curry. The three-ton rig was dragged up the fell on a sledge affair by a self-acting winch, with a 50 to 1 reduction gear, this having been modified from an old Morris Oxford car. This piece of ingenuity was the invention of the mine engineer, Harold Oglethorpe (shades of Captain W Borlase). The operation was not really a success and the men hated it. There was no road and the fuel for the wretched contraption had to be manhandled up hundreds of feet of fell. One day there was a serious fuel leak and the whole lot burst into flames. Due to the lack of water on site, the rig was virtually destroyed and no further work was done. Glencoynedale Head is a fine place on a summer's day but in normal Lakeland weather it is bleak and miserable, with little or no shelter.

Old Bloomfield was still the mine manager and because of his position he used the mine employees as personal staff. On one occasion when his pride and joy, a yellow Morris Isis car, failed to start, Walter Burnett and Terrance Mortimer were sent to repair it. Later, Bloomfield found that the tyre pressure gauge was missing and the two men were hauled up to the office for questioning. Walter had heard enough and informed the manager in no uncertain terms that they were employed by the company and were not here to do "foreigners". He came to Bloomfield's notice again, a little while later. The local men used to walk up the mine or cadge lifts on anything that was going up. Tommy Hodgson, who lived in Halton Terrace, was nearly always last up the road and was usually given a lift by Walter. One morning, Tommy broke with tradition and set off early. Old Bloomfield was driving up the road and seeing Tommy, slowed down and informed him that unless he got a move on he would be late. Aware of his erratic time keeping, Bloomfield enquired as to how he managed normally to get to work on time. Just then, with a roar, Burnett burned by on the

Velocette and Tommy replied: "Normally on the back of that!"

Bloomfield was a stickler for time and once, having assembled everyone, he told all the men that in the future they were to be ready to go underground five minutes before the start of a shift. Harry Holliday, who was an insomniac and always arrived everywhere an hour early, chirped up from the back "That won't apply to me then!" Old Bloomfield wanted to know who the manager was around here, to which he received the reply "You are, and you're not making a very good job of it!"

On one particular New Year's Day, Bloomfield arrived at the mine to find only two men at work, Arnold Lewis and Clarry Jenkinson, the latter being much the worse for wear. On enquiring as to the whereabouts of the rest of the shift, he was told they were at the New Year's Day hunt. Amazed at this Bloomfield later gathered the men together and told them that the next New Year's Day was to be an unpaid holiday, which brought forth from someone the remark: "That's not a holiday, that's enforced poverty".

The flood of 1927 — the damaged bridge over Glenridding Beck, and villagers clearing up after a night of destruction.

Sweeping up the debris outside the post office on Glenridding's main road. Below: rock breaking in the fields as the cleaning up campaign continues.

Flood waters subsiding in Glenridding. The Millcrest Hotel is in the background, the road bridge in the centre of the picture.

Levio, an immigrant Austrian miner at Greenside during the 1920s.

Miners working up at Keppelcove, 1928. Left to right: Bill Shaw, Fred Dewis, underground foreman Bill "Primer" Richards, Oswald Comer, Eddie Shaw.

One of the waterwheels of Greenside Mine.

Stripped to the bone —the Greenside loco underwent many transformations in its time. Overleaf: back on the tracks and heading for the Lucy, circa 1930.

W Jenkinson aboard the Greenside loco, 1930. Overleaf: Thomas "Scan" Hind and the Joy Sullivan air compressor on Warsop's Crosscut, Lucy Tongue Level, 1940. Below: Scan in the Smith's Shaft cage at the shaft top on Warsop's Crosscut, 1940.

Ginger Thompson with the Holman hoist at the top of Murray Shaft, 1938. Below: Billie Sinkinson, Ernie Jamieson, Abe Pattinson and W Currie with the diamond boring machine in the wilds of Glencoynedale, circa 1950. Glencoynedale was an inhospitable place to work. One day the machine mysteriously caught fire and the job had to be abandoned.

Five likely lads — left to right: Andrew Wilson, Joe Hope, Robert Jenkinson, "Danny" George Wilson, Alfie Brown, circa 1950. Below: the rescue team; Hughie Taylor (Greenside), unknown (Long Meg Mine), Henry Richards (Greenside), "Loll" (Long Meg Mine); front row: James Bardsley (brigade leader, Greenside), Arnold Lewis (Greenside), 1953.

The Greenside Mine in 1962. The Lucy Level entrance is at the extreme right of the picture. (photo, W Burnett)

The Lucy Adit before the mine's closure in 1962. Note the heavy cables running to the underground electrical gear and the absence of overhead wires for the old loco.

Top of the connecting shaft between the Low Horse Level and the High Horse Level. The valve was installed to direct water from the flooded levels of the High Horse to the hydraulic engine at Willie Shaft. The remains of one of the High Horse dams can be seen in the background. (photo, Richard Bird)

One of the many rises between the Low Horse Level and the Lucy Tongue Level. This rise formed part of the escape route to Glencoyne and provided access to the Alma Stope. (photo, Richard Bird)

The miners' escape route as it passes through the Alma workings, high above the Alma Level. Here the hade of the Greenside Vein is particularly evident. (photo, Richard Bird)

The vast and awesome Alma Stope, deep under Greenside mountain. Here the void plunges without interruption several hundred feet to the hoppers of the Lucy Tongue Level. (photo, Richard Bird)

Concrete beds for the compressors and electrical equipment in Warsop's Crosscut on the Lucy Tongue Level. (photo, Richard Bird)

The chamber in Warsop's Crosscut where the headframe for Smith's Shaft once stood. The shaft itself is under the corrugated sheets in the middle of the level. (photo, Richard Bird)

Rebuilt and under steam — the Sentinel in the carpark at Glenridding. (photo, Ian Tyler)

9 Dark Days

Monday the 7th of July 1952 was to be the blackest day in the history of Greenside Mine. The workings had now reached their lowest depth of 237 fathoms. The morning was beautiful and warm, the sun shining in a cloudless, blue sky as the miners made their way up to the mine from Glenridding. Among them were Leo Mulryan, Dick Mallinson, Johnny Miller, George Gibson, F Wynn, G Brown, Danny Wilson, Fred Dawes, Joe Hope, Clarry Jenkinson, H Richardson, T Hodgson, J Barnsley, Walter Burnett, Tom Hind, D Hodgson, A Cooper, Bill Murray, Edward Pool and J Hamilton. Along with many others they ambled up the mine road, the warm sun on their backs. Tragically, the day was soon to be cut short for some, for within the space of a few hours four of these men would be dead, several would be injured and some would be heroes. For all the survivors this would be a day they would never forget for the rest of their lives, for unknown to them as they walked along the road, a fire had been raging underground for hours.

The worst thing in any confined space is fire, for it uses up the oxygen, leaving behind only poisonous, deadly, and odourless fumes.

The seat of the fire was in the North Shaft which had been built by the Canadian Engineers during the 1940s. The shaft was a steel girder construction with timber inlay boards, and within the construction was a high tension cable supplying power to the hoists and pumps in the lowest workings. On the day before, Sunday, there had been a violent electrical storm, and it was later thought that during the storm, lightning had struck the cables on the surface. The charge had run along the cable underground and arced onto the steel framework, the sparks and flash setting alight the wood and other debris lying around the 200 Fathom Level. The shaft, acting like a huge flue, drew the fire into an inferno. Another possibility was that someone had left a lit carbide lamp on station, which could have flared and set light to something. Whatever the cause, there was a fire belching black smoke along the lower levels and sucking precious oxygen from the air up the shafts. Much of the smoke had by now dissipated leaving only the invisible, deadly carbon monoxide gas.

Unaware of the terrible scenario below, the men entered the mine to begin their shift and Leo Mulryan and Dick Mallinson set off down the 175 Fathom Level to the 940 Winze. The winding gear here was normally operated by Bill "Twig" Brown and his spannerman Wilf Jenkinson, and was affectionately known as "Twig's Spot". From here the men would descend to the 200 Fathom Level. Following behind were Johny Miller and George Gibson who were heading for the North Shaft. Arriving at the 940 Winze, Leo Mulryan climbed into a kibble and was lowered down by Bill Murray. Unknown to Bill the airline for the hoists had been damaged by the fire, a fact that was now registering on the dials of the compressor room up on the Lucy Level which were recording low pressure. Here Tom "Scan" Hind, realising that something was amiss, shut down the plant.

Before the plant was shut down, much compressed air had already escaped through the damaged airlines and this in turn had pushed the deadly gas through the lower sections where it was swirling, unseen along the 200 Fathom Level and into the bottom area of the shaft into which Leo was now descending. Because there was no power to control the air winch, the kibble descended on inertia and, of course, there was no power for return. Bill Murray heard Leo shouting to bring him up, but with the lack of air this was impossible. The air conditioning fans were also off, but in the space of time in which they had initially been on, the poisonous gas had consequently been pushed into other areas, and already dozens of other miners were being affected.

Leo was now in a bad way and so Dick Mallinson set off down the ladders of the 940 Winze in a lone rescue attempt. Leo's cries for help were becoming weaker through oxygen deficiency brought about by the carbon monoxide. The exertions of Dick Mallinson, in his urgent descent of the ladders, brought him to a state of near exhaustion. He managed to call out: "I'm coming up, it's just about got me!" but he was now so overcome, he succumbed and collapsed. Bill Murray called down desperately to them to hang on while he got help, and set off down the 175 Fathom Level. At Murray's Shaft he met F Wynne, J Brown and Jack Hamilton, the mine foreman, who were hurrying towards him having heard the commotion. Bill Murray, his lungs tortured by the gas, gasped out what had happened and then passed out. Jack Hamilton and others were now realising the seriousness of the situation and started to evacuate the men as quickly as possible. Eddie Pool arrived and took up station at the top of the 940 Winze, while Johnny Miller and George Gibson started down the ladders of the shaft in a brave last attempt to rescue their two mates. Alas, they both suffocated whilst on the ladders.

Tom Hind had already raised the alarm by telephone from the compressor room and rescue teams were on the way from the West Cumberland Coal Rescue Unit. Someone had also rushed back down to the village to collect the old army respirators which people still had from the war years. In the village, as the mines rescue unit came speeding towards the mine the radios, heard through open windows on this bright sunny morning, went off one by one. Fearfully, the villagers waited for news.

At the mine, the manager Mr C Conner, Walter Burnett and A Cooper had arrived at the shaft top where Eddie Pool was trying in vain to raise the kibble. His exposure to the gas was beginning to take effect and he too was overcome. Douglas Hodgson and Walter Burnett carried and dragged Eddie to safety. Climbing into the cage with others, Douglas collapsed, his feet hanging outside the cage. This went unnoticed and they were horribly mutilated as the cage sped upwards from the disaster area. Down below, Mr Conner and Cooper shouted and screamed down the shaft, but now all was silent. They stayed on station for as long as they could, but their respirators were archaic affairs. Soon they too began to feel the effects of the fumes and reluctantly started to pull back. Nearing Murray's Shaft they were relieved to see the members of the mines rescue, who escorted them to safety.

Miners Henry Richards and Clarry Jenkinson had both been involved in the rescue and were now themselves overcome. They revived and then both passed out a second time. J Barnsley was found in the seat of his loco slumped over the controls. Outside the Lucy Level, men were being brought out in mine tubs and cared for in the warm sunshine by Dr Armstrong, ably assisted by Nurse Arnison. The West Cumberland Coal Rescue Unit was also assisted by teams from Winscale and Kirkby Thore. The Greenside men themselves had been hindered during their evacuation of the mine by the inadequate old army respirators.

Due to the dreadful conditions which now prevailed under ground, all hope of saving the four men was reluctantly given up and the mine given time to clear. In fact, it was over a week before the bodies could be brought to the surface by rescue teams. The dead were all ex-military men and they were buried at St Patrick's Church, Patterdale, with full military honours. The Bishop of Carlisle officiated and Major J Cropper, the Lord Lieutenant, represented the Queen. This was the saddest day the tiny village and its inhabitants had had to endure. The small church and graveyard were packed as the four were laid to rest.

John Miller left a widow and two children; Richard Mallinson, a widow and four children; Leo Mulryan, a widow and a son; George Gibson, unmarried, was only thirty-five years old at the time of his death. He left a sister.

Shortly after the tragedy a local committee set up a disaster fund with the trade union and the Basinghall Mining Company for the dependents of the men who had been killed. The Greenside Fund, as it became known, was well supported. The Carnegie Hero Trust contributed and a special donation was given to Douglas Hodgson, who had suffered terrible leg injuries.

It was some time before the heroism of the rescuers came to public notice, but their gallantry was duly recognised when, on the 22nd of August 1953 at Glenridding, awards were made by Major J Cropper, Lord Lieutenant of Westmorland.

The rescue teams had worked bravely in what was to them an alien environment, and two members of the team (*see below) received awards for rescuing miners from the interior of the mine at the time of the accident. They were both from the Winscale Rescue Station.

Roll of Honour

Mr Cyril Connor	Manager	MBE
Mr Walter Burnett	Electrician	BEM
Mr Edward Pool		Letter of Commendation
Mr T Hodgson		Letter of Commendation
Mr F Wynn		Letter of Commendation
Mr A Cooper		
Mr J Brown		
Mr John Miller	Posthumous	Edward Medal
Mr Richard Mallison	Posthumous	Edward Medal
Mr William Gibson	Posthumous	Edward Medal
Mr Richard Mallinson	Posthumous	Carnegie Hero Certificate
Mr John Miller	Posthumous	Carnegie Hero Certificate
Mr William Gibson	Posthumous	Carnegie Hero Certificate
Mr Douglas Hodgson	Survived	Carnegie Hero Certificate
Mr W Kirk*	Station Officer	Queen's Commendation
Mr R H Glaister*	Brigademan	Queen's Commendation

It was shortly after the incident that certain radical changes were made at the mine. Carbide lamps were to be replaced with Ceag electric lamps, though these were generally not liked. They were heavy and did not give an overall glow like the friendly carbide, and two men coming face to face blinded each other. Possibly worst of all, you couldn't light a "Woody" by electric light.

It is believed that the shaft hoists on the 940 Winze and the North Shaft were converted from air to electric winding.

This accident made the Mines Inspectorate realise that Greenside had no secondary exit, all comings and goings being made by the mile long Lucy Level. This was considered totally unacceptable, and they insisted that a secondary exit was effected without delay.

A first aid and rescue team was to be formed, and James Bardsley was elected as leader. Arnold Lewis, who was trained in first aid, was also to be on the team. Arnold's brother-in-law George Gibson had been lost in the disaster, and Arnold would himself have been on duty but he was off work at the time. Another team member was to be Thomas Firth.

For training purposes the new team were given every facility and an actual training tunnel was built to simulate conditions underground. With guidance from the NCB team in Newcastle, training sessions were carried out every two weeks. New underground breathing apparatus was supplied, known as Savox and made by Siebe Gorman & Co, the sets having a duration of forty-five minutes. These were much lighter and provided far easier movement through the manways. Although this system was the best available at the time, it was not liked by the team. First aid stations were set up underground at the 175 Fathom Level and at Smith's Shaft top. The rescue team was also supplied with the Oldham electric lamps.

A new code of safety practice was introduced, perhaps the most significant move being that the mine was inspected one hour before the commencement of each shift by a foreman and timberman, providing the workings had been vacated for fifteen hours or more.

Feelings of despondency and apathy lingered on at the mine, for the tragedy was not easily forgotten. Miners anywhere are a tight knit community, and ranks were closed against the prying of those inquisitive people and members of the press who came to satisfy a morbid curiosity.

Before the emergency exit was commenced, a thorough examination of the Low Horse Level was made by Jack Blair, an underground manager. This was to see if a way could be secured through the old workings and a connection made with the Lucy below. After a thorough investigation it was confirmed that it was impossible to break a way through, for the whole system had collapsed with many huge blocks in an unstable condition. It was considered more prudent to commence a new ladderway rise to ensure a safe escape route. The driving of the shaft was completed in double quick time, as two of the best driving men had been put on the job, Fred Dawes and Danny Wilson, and the ladders were put in by Joe Hope.

The rise was put up from the Alma stope to the Low Horse, and from here up to the High Horse an old shaft was used which had brought the pipes down for the hydraulic engines years before. The final breakthrough into the High Horse Level was spectacular, no one having been in the old workings for years. The dam constructed for the retention of water for the hydraulic engines proved a formidable obstacle, as Fred and Danny burnt out the tungsten bits attempting to drill into the reinforced masonry. Here the air was poor in the confined space of the sluice chamber, and the journey each day entailed a ladder climb of around seven-hundred feet.

Finally they broke through the dam not knowing how much water still remained. There was a rush of air and a fair amount of water and then a wind howling through the hole, causing waves. Elated, the miners carved their names at the shaft top for posterity, and the date, 1953. The mine was safer than it had been, and it was probably a relief to many of the men working there to know that they now had that long overdue second exit.

By 1954 things were back to normal. Miners from Penrith were still being brought by wagon driven by Isaac "Ike" Bennett, while the men from Threlkeld drove up in style by bus. Because of its size and rock strata, the mine was selected to have its own drill testing section, which came under the Ministry of Power and was run on site by Mr Broomhead BSc, ably assisted by miner Henry Richards. The section had numerous tasks to perform, including testing the various types of drill bits, analysing the quantity of dust produced both in wet and dry drilling, and also what effect this had on the miners by way of silicosis phthisis.

After each test the sealed area had to be washed out so that it was completely free of dust and accurate readings could be taken for subsequent tests. In order to ensure a regular supply of drill steels at the working face, a drill-doctor station was set up on the 175 Fathom Level. Here the tungsten bits were re-sharpened, saving the many man hours used in having to take the steels out to day. The work was done on an Atlas LSB 62 grinder by Henry Richards and Percy Dewis. Prior to 1954 most of the rock drills used were of American manufacture, but these were costly and subsequently replaced after a year's trial with the new lightweight Atlas Diesel Co Ltd RH 656-2W rock drill with BMK 61-A2 pusher leg and Sandvik Coromant tungsten-carbide tipped drill steel. With these new drills it was estimated that a 7 foot by 7 foot drift could be driven forward 95 feet per month, with an Eimco for mucking out. This was an increase of 35 feet over the previous method.

The miners were firing ten rounds every eleven days, where previously only six could be achieved. The explosive was brought into the mine by two special mine tubs of wooden construction, and as much as 250 pounds was stored in the underground magazine, as anything up to 500 pounds could be used every day.

Production had steadily grown since 1947 and by 1954 the plant was processing nearly 26,000 tons of ore, the concentrates being sent to Newcastle upon Tyne by road instead of rail. A local firm, Mandals of Greystoke, did most of the carrying, driver Tommy Charlton delivering about twelve tons at a time. If further or larger deliveries were required, the job invariably went to Border Wagons of Carlisle.

Tommy Hind was still in charge of his beloved compressor, but she was getting old now and was not so reliable. One day during 1954 when "Scan" switched on there was a crackle, a sizzle and a huge blue flash as condensation had formed under the shiny black covers. Walter Burnett came to see if the job could be sorted, and an hour later Mr Flynn, now a senior manager, appeared with numerous plans and diagrams and promptly took over. The mine manager Mr Conner was next on the scene to check up on the progress and to be reassured by Mr Flynn that there was no problem. Walter smiled to himself and thought about the three hours it had taken just to get the cover off. Time rolled by, and it was gone six o'clock and they were really no further. Then the distant rumbling of the loco was heard, although the last shift had gone ages ago.

Walter and Flynn climbed down off the compressor and went out into the level, and it was indeed the loco approaching. Aboard were Mr Conner with Mr George Bussey, the chief electrician from Thornton's Electric of Wrexham and his apprentice, the boss's son John Thornton. Mr Conner, realising assistance may have been required, had telephoned Wrexham, and Thorntons had sent the men up post haste. They got to work and identified the fault, and had the machine ready and operational for the following day. Mr Flynn was incensed at outsiders being brought in to repair his machinery. Muttering all the way out to day his final words were the usual "I'll see the GM about this". Walter gave a wry smile and thought about his sixty-eight electric motors that broke down at the rate of at least one a day — if it wasn't a hoist it was a conveyer, and so it went on. His most difficult or most hated job was when a tub jumped the rails and caused damage to the big three-inch cable which fed power into the mine. This took all day to resplice, which was more often than not the case, and entailed using sylvester wrenches to support and hold the huge cable whilst the intricate job of repairing it took place.

The new escape route was now official and had been passed by the Mines Inspectorate. The route was checked regularly and this was normally Jack Blair's job. It was a long haul up through the system, even from Lucy Level the climb is 760 feet on vertical ladders. It was tough training for the newly-formed rescue team, who had to climb from the 237 Fathom Level to Lucy Level (1,422 feet) then another 760 feet to Glencoyne adit, and in full rescue kit plus breathing apparatus weighing around 40 pounds.

Eddie Pool, a fitter, was keen on the mine's history but had never seen the upper workings and, knowing his father had worked these areas in times gone by, asked Jack if he could come along for company one night. This was arranged and they agreed that they would walk back across the fell. Two or three hours elapsed before the two men returned to hand in their lamps, where they found Harold Oglethorpe waiting and none too pleased. Eddie, for once in an exemplary career, had forgotten to tell anyone where he was going, and as a result a search of the mine was about to be started. No doubt it had been Tommy Wilson the lamp man, a disabled war veteran, who had noticed that the lamp was missing and advised Mr Oglethorpe.

It was around 1955 that a spectacular accident happened topside when the Siemann's crusher was being overhauled and serviced. The work was being supervised by Harold Oglethorpe and Tommy Hodgson was applying molten metal to the crusher base. The metal suddenly exploded and a fire started setting the conveyer belt from the crusher alight. The force of the blast blew the two men off the crusher and both sustained injuries requiring a spell in hospital, but happily both survived to tell the tale.

Production at the mine was falling but investment in new machinery was ongoing, and to make the operation more efficient a new Eimco Model 12B rocker shovel was ordered on the 25th of July 1956, and delivered on the 9th of September at a cost of £1,135. These wonderful little machines were invented by John S Finlay and E B Royle of America, and were first used in the North Lily Mine, Eureka, Utah in 1931. The invention was first known as the Eimco Finlay Loader, and this made the backbreaking job of mucking out by hand as obsolete as hand drilling.

In the search for more ore, new headings were still being driven. Henry Richards was a miner of many years experience, the son of a Cornish miner and senior driller, and had worked at Greenside since leaving school in 1931. Working in a very wet area, Henry and his mate had drilled the face, set the charges and retired to a safe area. The charges went off one after the other and the count went on.....one short. Henry went to check, and as he approached the forehead there was a deafening bang and yellow and orange flame belched out. The blast lifted Henry off his feet throwing him unconscious down the level. Somehow he survived the full force of the blast, and after a time recovered, but not without sustaining injuries to his eyes and ears.

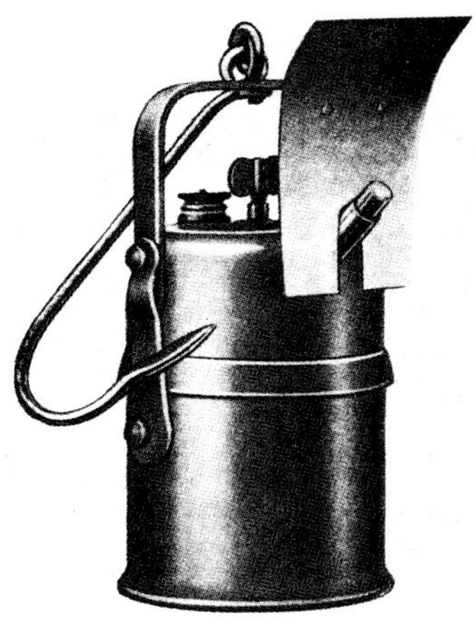

Kaffir Lamp Fig. 63

MINING LAMP

fitted with

ECCENTRIC LOCK AND
OVERHANGING REFLECTOR
FOR DAMP MINES

One size in Steel (tinned after manufacture) with
Brass Water Vessel Base

Specification as Fig. 60 No. 1

Code Word KAFFA

10 Operation Orpheus

By 1959 the geological reports were disappointing and it appeared that the mine had little future. In fact, the miners themselves could see the writing on the wall and many were leaving to secure employment elsewhere. But then a new lease of life presented itself.

This strange rescue package came from a government department known as AWRE (Atomic Weapon Research Establishment) at Aldermaston. They in fact took over the mine under their officer commanding, Mr E R Drake Seager, and his staff of thirty. The project was to simulate an atomic explosion underground. It was the word atomic which spread round the area like wildfire and rumours of explosions and radiation were rife. Basically, the idea was to explode two separate charges, one of 250 lbs of TNT linked to a further one of 500 lbs. The explosion would be controlled in a blind heading for maximum effect, and seismic readings of the shock waves would be taken both underground and on surface to record the results.

The area chosen was on the 175 Fathom Level at the southern end of an eastern crosscut, directly under Sheffield Pike. This was not the deepest area of the mine but it was a totally barren area, and was easily accessible from Murray's Shaft. It took months to prepare the area, in fact extra people were employed for the job. Two chambers were made ready, one for each of the charges and both were wired up, the idea being that the main charge of 500 lbs would be fired, and the second charge was set to act as a baffle charge.

On the surface, a police control point was built at the entrance to Lucy Level to ensure only authorised personnel entered the mine. All people connected with the operation carried self-rescue packs. The day of the detonation commenced with a countdown from 0900 hrs to firing at 1900 hrs with no less than thirty-one countdown sections, directly involving at least twenty mine and AWRE personnel.

After the briefing on the day of the firing the team was taken up the level by loco driven by Mr Wilson, the mine manager. Mr Corlett was in charge of mine staff and Mr Drake Seager in charge of the AWRE staff. By 1230 hrs the charges were ready and the mine totally sealed and checked. The party returned, the charge was armed and the mine evacuated. All personnel were in the fire control room at 1830 hrs. The last countdown started, and on the final pip of the 1900 hrs BBC time signal the charge was detonated. It was something of anticlimax when the main charge went off, and the second failed. Later, it was discovered that this was due to the fact that the ferocity of the first explosion had in fact destroyed the connecting wires.

The charges had been laid in a totally sealed area, sandbagged floor to ceiling, with timber bulkheads and steel doors. Everything was done that could be done to contain the blast within the confines of the chambers. Two days elapsed before anyone was allowed to enter the mine, and the first to enter were escorted by men from the Kirkby Thore Mine Rescue Team. The mine was first tested for gas from the 237 Fathom Level right through to Glencoyne Level, the air levels found to be satisfactory. The mines inspectors passed the mine fit for working.

Operation Orpheus

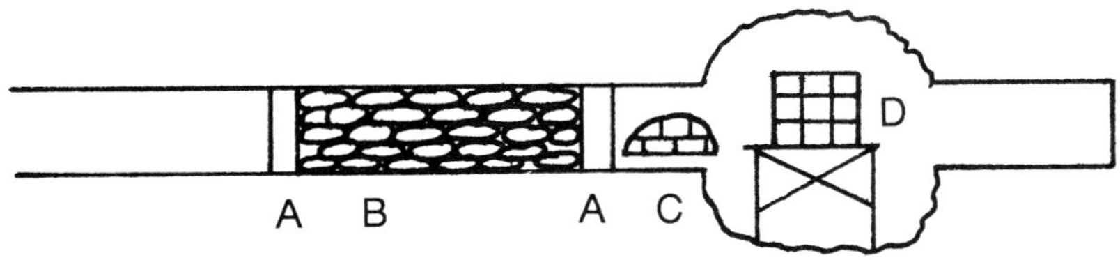

A steel doors and brick walls
B sand bags
C 250lbs TNT
D 500lbs TNT

170 FM south

The first job was to clear the debris away and remove the 250 lbs of explosive.

Work started in 1960 on the rebuilding of the damaged areas in preparation for another experiment. Two of the men on the job were William Sinkinson and Alex Santamera. In the 175 Fathom Level the mine was dry and warm and the men were building sandbag walls. During one lunch break, William and Alex got permission to go and look for quartz crystals but were told not to go beyond the North Shaft.

It was Arnold Lewis who realised that they had not returned, and they had certainly been gone too long. Fred Dawes immediately said he would go to look for them and Arnold volunteered to go with him. They began checking the side drives, and down one of these was the ironically named Dawes Stope. It was in here the two men were found. The experience of the two searchers told them as soon as they entered that the air was not good. Being so far down and in a blind heading, the area had not ventilated and the thought of gas was uppermost in their minds. On climbing into the stope, their lamps revealed the motionless forms of the two men. Alex was in the bottom of the stope and William could be seen up near the top on some ladders.

Arnold Lewis, being a first aid man and member of the rescue team, told Fred to fill his lungs with air and they would try and get Alex out first. Both men lunged across the stope and dragged Alex across the rough ground to the top of the ladderway where the air was fresher, in a bid to save him. The shortage of air coupled with their exertions left the two men gasping as they got to the access ladder, and with great difficulty they managed to get down it. From here they stumbled down the level where they met the foreman and warned him of the gas. The next thing Arnold knew was a light flashing in his eyes as he regained consciousness, and looking round he could just see Fred. They were both taken up and out to day.

Later, they found that their rescue attempt had been in vain and that both men were dead before Arnold and Fred had got to them. However, their brave and unselfish rescue attempt did not go unnoticed by the authorities, and both were awarded the Queen's Commendation for Brave Conduct. It was very shortly after this accident that the AWRE left Greenside, in August 1960, as did Arnold Lewis who went to work in Kendal, where he still lives today.

On the departure of the AWRE the mine resumed working normally, but the break in production had been too long. Although the plant was in first class order it was taking some time to get under way again.

The exploration work in the lower area of the mine had hit shale. Two miners who drilled this section insist to this day that it was oil shale, but it made no difference, and during April 1961 the last tub of ore was drawn from the Lucy Level. It was now the sad duty of Mr H E Evans, the administrative agent, to commence the unenviable job of clearing the site and winding up the company's assets both underground and on surface.

To assist Mr Evans in his task twenty-five men were retained, many of the original workforce having sought other jobs in the area. Places like Kirkstone quarries, the Eden Valley quarries, Shap quarries and Threlkeld Granite were able to absorb some of the skilled men, whilst others turned their hands to other trades.

The auction of the plant was carried out by Penrith Farmers & Kidds and any remaining machinery was cut down and sold for scrap. One of the largest and heaviest items was the crusher, which was lifted by a huge mobile crane and transported to Threlkeld Granite Quarry. Loco No.4812 was sold to McKechnie's where it worked at Force Crag Mine, Braithwaite. The total bill at the sale was £695.8s.11d. Loco No.1192 was sold with two batteries to the Weardale Lead Company, Durham, who paid £600. They also purchased a number of pumps.

Loco No.2489 was sold to McKechnie's at Sandbeds Mine, Caldbeck. Either No.4472 or No.1215 may have been purchased by Goodycrane Quarry, St Austell, Cornwall, who offered £250. They did purchase some of the 16 cubic-feet tubs for £15. One of the Eimcos went to La Porte's Silverband Mine. Many of the newer Hudson tubs were sold for £15, others for as little as £8. Some went to Force Crag, some to Sandbeds, which were both operated by McKechnies under Bill Shaw. P Creggain of Maryport purchased a weighbridge and overhead crane, Blencow Lime of Penrith purchased a quantity of piping and ladders, and Associated Lead Manufacturers purchased one of the large compressors. R R Britton Ltd, of Bristol, purchased the ball mill, and George Cohen & Son of Shepherds Bush took the Niagra screens for £160 and the Holman compressor for £275. Muirshields Barytes Mine at Renfrew purchased many of the mining lamps, and Barking Metal Trading Co of Essex bought nearly all the electric motors. Other mines purchasing items were Closehouse Mine, Durham, and Long Meg Anhydrite Mine, Langwathby, near Penrith. Shortly afterwards the prefabricated buildings were removed and any buildings regarded as unstable or unsightly, regardless of historic value or interest, were bulldozed and the remains covered with earth.

The waste heaps had long been a problem, ever since the mine had been pumping the fines on the old dumps. The company had tried to stabilise and cultivate the area to disguise its obvious outline. A labyrinth of concrete drains and wooden drainage columns set inside the dumps were installed to stabilise them and so prevent the fines from washing away. The first efforts were made as early as 1940 and continued until 1948, when no less than 20,000 tufts of marram grass were planted. Now and again a liberal dumping of manure was spread to help bed the fines down. There was even an electric fence put round the dumps in 1948 to stop sheep disturbing the herbaceous plot. Experts were called in during 1952 to assess the situation, and seed and fertiliser recommended but all to no avail. In 1953 a further batch of sewage sludge was pumped on top and a new six foot post and wire fence was erected.

By 1962, when the mine had closed, the legacy of something like a million tons of sterile dross had been dumped, of which nearly 70% was silica, and 12% alumina plus zinc, magnesia and ferric oxide. Not exactly the finest medium for cultivation and, eventually, nature would do the job anyway.

In the late 1980s an enormous landslip virtually swamped the old smelter building with tons of fines, and to the rescue came ex-miner Eddie Pool with his JCB and removed tons of waste. Consequently, more action was taken using stakes, sawdust and wood bark. By 1990 the rabbits were burrowing madly, and the National Park continues to spray liquid manure and grass seed, which is fine until it rains!

When the time came to finally secure the mine itself, the levels were either blown up or pulled in, but the Glencoyne Level, for some reason, was bricked up, the work done by W Curry. Drainage pipes were put in both Lucy Level and Low Horse. All the dams were breached so that the water returned to its original levels, and streams which had been used to supply leats were diverted back to their original courses. The buildings still standing now are the joiners' shop, offices, Bell Cottage, stables, mine hostel and part of the old smelter which was used in its later years as the electrical shop and stores. These have been converted for other purposes and bear no architectural resemblance to the original mine buildings.

Once all the activity of the closure had died down the valley became quiet once more, but it was greatly changed. Gone was a complete way of life for many people. Some men had worked at the mine all their working lives as had their forefathers, brothers, sons and friends, and this was a sad passing. The small churchyard reveals many memories on the headstones, some of heroes, some victims, and others who came to rest here.

High above in the valley there is also another epitaph and memorial in the huge spoil heaps of Lucy Level, the Low Horse and the High Horse, mountains of stone having been wrested from the earth by basic physical effort. It is difficult to imagine the hardship and the toil, the privations and the danger which these hardy men endured in the cavernous workings of Greenside Mine for almost 200 years. In this time, over a million tons of ore were crushed, and a vertical hole in the region of 2,750 feet now extends from the uppermost working to the lowest level. The area of the ore body has been followed for miles and the miners, over the duration, hacked and blasted no less than twelve-and-a-half miles of levels, not to mention at least 11,750 feet of main shafts and rises. It is true to say that nearly half of this was done by sheer physical manpower.

Beneath the ground all is still and quiet, apart from the dripping of water. The 760 feet of escape ladders from Lucy Level to Glencoyne Level hang rusting on the shaft wall, the odd tub is to be found in the Alma stope and on the 36 Fathom Level. Only a few fit and keen mine explorers are privy to the whitewashed area of Smith's Shaft and the compressor room with its collection of artefacts. Clear blue water now fills Smith's Shaft to the collar. It is 1,460 feet deep and below it is the true bottom of the mine. All of this lower area is now totally flooded. Here is the place where hundreds of men worked, where acts of bravery occurred, where men died. It is very dark and the silence is eerie, and the spirit of Greenside Mine lays quiet under its rolling green top.

11 Kirkstone Pass & Caiston Glen Mines

Numerous small trials are to be found within the Patterdale valley. The workings of the Kirkstone operation are hidden away high on the east and west slopes of the pass, but the dark grey spoil visible on the west side of the road is from an old shaft working which was sunk fifteen fathoms onto a very large quartz vein some six feet wide at this point. This is an east-west vein which cuts across the valley and bears a little chalcopyrite. It seems that, during the year of 1820, the landowner, Lord Lonsdale, ordered this old shaft to be filled in, presumably for reasons of safety, and no further details are to be found regarding the working.

Above the shaft are the remains of the miners' hut which is also sited by the old packhorse road. Four-hundred feet directly above the miners' hut are to be found the main workings. It is evident that the miners originally excavated the vein on surface and then drove the adit, which is directly behind an impressive cascade of water. The level was hand-drilled and extends some forty-one yards to a blind heading. Above this working is an extensive surface trial which has been taken down on the vein some twenty feet and which floods badly in inclement weather. However, no connection was made with the adit working below. At this point the vein shifts its direction to north-west to cut right through Middle Dodd and then appear in the secluded and quiet valley known as Caiston Glen.

Looking across to Caudale Moor one can see the milky white quartz vein striking up the fell, and here, nestling in a profusion of broken quartz, is to be found a very small trial and dressing floor. All these workings were excavated between 1820 and 1860 at the latest, and it appears that some hæmatite as well as copper was taken, but in very small amounts.

In Caiston Glen, under the looming western shoulder of Middle Dodd, lies a small working which consists of a short level cut into the solid rock. The vein can be clearly seen in Caiston Beck which bubbles on its way close by. Directly above, a small sump has been sunk on the vein but this does not connect with the workings below. Higher up the fell and about fifty yards to the south east is a grassed-over surface trial. The level is of a reasonable size, being about 4 feet wide and 5ft 9ins high, hand-drilled, and probably driven around the 1870s.

12 Myers Head

Just past the Brothers Water Hotel, and to the west, lies the delightful valley of Dovedale. With its leafy trees caressing the hillsides, the valley is divided by the crashing cascades of Hartsop Beck, once the supplier of water to the now derelict Hartsop Lead Mine. The head of the valley is barricaded by the lofty ramparts of Dove Crag, a famous haunt of rock climbers, while to the south is the magnificent Hogget Gill, containing a spectacular series of waterfalls and where modern-day gold panners have, supposedly, found minute specks of this precious metal.

The twisting road soon passes the oddly shaped Brothers Water and, nearby, the ancient village of Low Hartsop where many of the buildings date back to the 1500s. The reason for the early settlement could be its sheltered setting and the steady supply of water from the streams of Hayeswater Gill and Pasture Beck, ensuring the lush grass in the valley. Or perhaps the mineral veins dissecting the dominating bulk of Hartsop Dodd, looming above the tiny hamlet, enticed those early settlers to this site. Certainly in very early days this was the most densely populated part of the Ullswater valley.

The geology of the sett indicates three veins which, for ease, we will call A, B and C. The main vein, A, courses through the Borrowdale series in a NNW to SSE direction and on which the main shaft working has been sunk. South of here, the vein is exposed in the stream called Pasture Beck, and has been tried by the old men by surface pits and, further south, by two stone-arched levels. The vein contains galena with a silver content of only sixteen ounces per ton, much less than many other Lake District mines. Also to be found here is zinc blende.

The vein B, coursing east to west, is a lead vein. This vein is exposed at surface just below Dodd End Mine.

The other vein C, hades north-east to south-west and has been worked by two trials, one directly on the vein which bears a quantity of zinc blende, and at another point where a crosscut has been driven.

The site is quite extensive, the oldest workings being the ancient pits near to the point where vein A crosses the gill and where the old men first tried for lead. The two old stone-arched levels have been driven just to the east of Pasture Beck. A little lead is to be found on the tips. These trials are both subsequent workings to the pits and were made in the early 1800s. Unfortunately, both these workings are run in after a short distance.

The open works high on the north face of Hartsop Dodd would have been made around this period. The vein here is totally exposed through to surface and would have provided easy pickings for the old men. This travels in a north-east to south-west direction and is nothing to do with the Myers Head Mine, but belongs to the Dodd End Mine workings.

The main enterprise came much later and was on a grand scale, and the surface remains are much in evidence today. During the early 1860s the Low Hartsop Mining Company was formed by Mr Coultas Dodsworth of Haydon Bridge, Mr George Head-Head of Carlisle and a Mr Jameson, these last two both being directors of the Greenside Mining Company. Flushed with their success at Greenside, they possibly thought that this independent venture would secure more wealth. Mining is, to say the least, a risky business and all was not to be plain sailing. It was decided that a shaft would be sunk to prove the vein at depth and this was indeed a rare decision in Lake District mining, for few shaft workings could boast success. Most mines were drift operations which were generally cheaper to operate being self draining and entailing no winding costs. Despite these precedents, however, the shaft was commenced around 1866 on the west side of Pasture Beck, adjacent to the site of the old pits. Initially, a horse whim was erected for the removal of the waste rock and water, though this was only a temporary measure until a more efficient waterwheel and pumps could be erected.

By the end of 1867 the shaft was down to the required depth of thirty fathoms, and a crosscut of twelve fathoms was being driven east to the vein. Reports dated April 1868 were sent to the owners informing them that water was presenting a major problem. The vein having been cut resulted in some lead ore being won and this was ready to be drawn out, but the water in the shaft bottom was already causing difficulties, and it was essential that pumps were installed to keep the mine workable.

Evidence of this massive project, which must have been completed in record time, can still be seen near to the old smithy, which is situated on the west bank of Hayeswater Gill. Here the miners built a wooden dam with a cobbled run off, and on the eastern side water was diverted via a sluice onto huge wooden launders which were supported by eleven stone stanchions. These carried the water the 150 or so yards from the dam to the wheel, which in turn drove the drive rods spanning the beck to the shaft. Here, a balance bob converted the horizontal strokes to vertical energy for the pumps. According to the reports this work was completed by November 1868.

The conditions underground were horrendous, the unstable nature of the ground requiring the miners to resort to bricking and cementing walls to support the rotten bedrock. It was the crosscut and part of the vein, particularly in the northern heading, which were the most unstable.

These painstaking efforts, however, did not totally resolve the problem, and in a report from the agent, George Harrison, Lord Lowther was informed of the slowness of progress. The good news was that the vein was showing lead and so it was decided to press on. By the

end of January 1869 the situation was becoming critical, for little ore was being raised to help fund the operation. The last straw was a major collapse, and the sandy vein material, mixed with an inrush of water, simply choked the pumps. As a result the operation came to a standstill and after much deliberation the project was scrapped.

Not far away, in Grisedale valley, the Eagle Crag Vein was being worked under the watchful eye of manager Richard Grenfell, and it was he who instigated the move of some of his workforce to the now derelict site of Myers Head early in 1870. The men were put to work clearing the shaft and the underground workings, and once this had been completed the work of mining began.

By the end of 1870 a total of three tons of lead ore had been raised, and with the more powerful Cornish pumps installed, the miners must have been somewhat comforted by the fact that the water level in the mine was being kept to a minimum with the wheel turning virtually twenty-four hours a day. It must have been quite a sobering thought that on arriving for work you had to descend 180 feet of ladders and then walk through an area of unstable ground for twelve fathoms to the vein, your workplace for the day, knowing that the only way out was back up the shaft.

The vein continued to be driven from the crosscut nineteen fathoms in a northerly direction, but this proved to be an awful place where the water poured in and the mineral was very poor. Consequently, it was decided to pursue the vein by driving southwards, and this course of action was taken for many months though with little return. They had now driven ninety-four fathoms south of the shaft crosscut and were cutting into a band of slate. Some seventy feet prior to this point a small quartz string had been cut and was tried for sixteen feet in a westerly direction, but nothing of importance was found.

The north and south vein had been an expensive project, taking years to do; little workable ore had been found, and what little was raised had been insufficient to cover costs. Understandably, the company was becoming anxious as its efforts were going unrewarded. The miners laboured on and the southern drive was pushed even further in the hope that the vein would improve. It was reported that during 1876 some ore was removed from the mine by "a flat cart pulled by a team of three horses adorned with gleaming brasses", but at best this could have only weighed two tons.

The pumps strained on, lifting gallons of unwanted water from the shaft bottom, as the clanking of the great thirty-foot wheel echoed in the still of the night. The constant use of the wheel meant regular repair work had to be carried out, and this was done normally by one George Harrison (no relation to the agent). In 1877, for repairs to the wheel and leats, he was paid £2.10s.0d for three day's work. Thomas Smith, another workman, purchased candles to the value of 2/11d, soap for 1/- and oil for 7d. James Harrison, who was possibly related to George, worked for two days for 6/6d, whilst George Morton received 5/6d for one day's work.

Mr Richard Grenfell was convinced that the mine could be successful, but the protection of the men's lives was of paramount importance, and he realised that not only was the vein creating much water, but also that it was coming in where the level passed under the stream. He drew up plans to divert the stream but he died before they could be initiated.

More repairs were carried out on the surface during June and December 1878. William Allcock, a joiner, was paid 13/9d for work done, and John Priestman was paid 9/- for carriage, most likely being the transportation of ore down to the smelter at Greenside.

Unfortunately, disaster befell the mine shortly after these reports were made, and it occurred whilst the miners were driving the north-south level forward. A fault in the vein was cut and gallons of water gushed into the mine. The miners ran for their lives, the only light provided by the tallow candles flickering and guttering faintly on the level walls, guiding their way to safety. The rising water forced them on through the flotsam of old powder kegs, timber, sacks and shovels, the debris hampering their progress as they struggled to reach the shaft bottom. There, lungs bursting with exertion, the men tore at the rungs, fear, panic and, above all, the will to live giving them the strength to climb the 180 feet of near-vertical ladderways as the swirling water rose rapidly behind them from the blackness below.

It is possibly the fact that the pumps were still operating, and one or two vital strokes managed to keep the water back for long enough, that allowed the men time to reach the surface safely and so avert a tragedy. The shaft was flooded to the collar and choked with debris — such was the strength of the flood. This was disastrous news for the company, which had not enjoyed the best of fortunes.

The mine was lost. So too was nearly all the mining equipment. What was left was auctioned off on the 11th of March 1884. The waterwheel, plant, tools and any remaining tackle were sold for a mere £85 to the Cumberland Lead Mines Ltd, the waterwheel being dismantled and erected at their Thornthwaite mine where it was in use until 1920. The Low Hartsop Mining Company was requested to remove all scrap from the site and make good by the 11th of March 1884. Due to this total, devastating loss after years of investment with little or no return, the closure was brought about not only of Myers Head but also Eagle Crag Mine.

On the northern bank of Hayeswater Gill are two trials, the lowest of them at OD 610 feet having a fine stone-arched level which has been driven north-east. It has cut a zinc-bearing vein which also contained a little lead. This level was driven just prior to the 1860s, and at this period zinc was regarded as a virtually useless mineral, and, no doubt because of this, the work was stopped.

Nearby, at OD 690 feet and a few yards to the east, lies the other level entrance, initially driven through boulder clay with the roof supported by timbers until it finally strikes solid ground. This, presumably, was an attempted crosscut to the vein, with the intention of connecting to the lower level and eventually stoping the ground out between the two. The supporting timbers have now gone and are merely stumps, and there is little sign of the vein. The date of working will be a similar period, around the 1860s. Certainly neither of these would have provided any return by way of ore produced.

It is interesting to note that the miners here built so many stone-arched levels when there were trees in abundance. It was, possibly, the influence of miners from Alston.

The huge bulk of Hartsop Dodd stands guard over the tiny hamlet of Low Hartsop, once a thriving little community with its stone cottages, some with spinning galleries and, as mentioned earlier, one of the oldest settlements in the valley. Adjacent to the two trials already mentioned, and in fact situated beside the road to Myers Head Mine, are to be found the ruins of a small corn mill, which no doubt supplied the hamlet with fresh flour. The remains of the leat which carried the water to drive the waterwheel can still be seen behind the mill, and the red sandstone millstones lie there, slowly being covered by the collapsing walls. The millstones are dated 1870, the height of the mining activity in the area, so possibly the mill was built to supply the extra requirements of this then thriving village.

13 Dodd End Mine

Deep in the heart of Hartsop Dodd lie the remains of Dodd End Mine, though unfortunately little of the history is known of this extensive working. The vein, containing galena, bears NW-SE and was originally tried by a shallow open cut at OD 1,100 feet. At thirty feet below this a level has been driven to prove the vein at depth. Midway along its length a rise has been put up to surface and breaks ground just in front of the open work. The vein at this horizon, however, is virtually non-existent.

These workings could be attributed to Thomas Dodd who was prospecting in the area in 1802 but who soon became disillusioned after only a short period of time. The small dimensions of the level would indicate that the trial was made around this time. Higher up the face to the south-east a pocket working, again directly on the vein, is to be found. From the valley below, the adit entrance looks truly impressive but in reality it extends only about six feet and was known locally as "Cruel John's Hole", the origins of this name apparently having been lost in the mists of time.

On the 14th of November 1871 a report was made to the Earl of Lonsdale informing him that a new vein was being followed in a north-south direction directly into the heart of Hartsop Dodd. This drive was to be known as the Low Level, starting on the north side of the fell. Near the entrance is a well-made hand dressing floor, thought to have been constructed by the Patterdale Mining Co, which was then working the Myers Head operation nearby. No doubt the company, having inspected the vein and the old trials of Thomas Dodd, decided to make a major trial. This entailed driving forward some 130 yards to come underneath the old workings and then connecting the two with a 200 foot shaft.

To start with a rise was driven upwards from the Low Level, whilst above a service adit was driven south to meet the old east-west level. After ten yards the connection was made, so winding gear was erected and a shaft was commenced to meet the men rising up from below. It is not known why such an expensive method of working was adopted, as it was usual for a shaft to be driven entirely from below. The men rising won the race, having driven upwards of 120 feet, but unfortunately the surveyor was off target by about six feet and the shafts did not meet. The company would have been anything but pleased about this as the shaft could not be used for direct hauling because of the kink. A considerable amount of work needed to be done to straighten out the shaft, and it appears that this work was indeed started but was never completed. Two small trial levels were driven off the shaft for a few feet but were not pursued. By now the Low Level had been driven a total of 282 yards.

It would be prudent to suggest that this working would have taken something in the region of six to seven years to complete, due to the sheer physical effort of hand-driving and blasting forward a level of this size. It begs the question: were the company's men acting in a wise and practical manner, or were their ideas and plans perhaps coloured so convincingly by dreams of another Greenside?

As to costs, the driving rate was say 70/- a fathom, shaft sinking about 120/- a fathom, and then there would be railing out, so the basic cost of this operation would have been approximately £1,700. On top of this would also have been the purchase of gunpowder and other ancillary items, and then timber, carriage and surveyors fees — an expensive venture for little or no ore sold, and with very little showing along the whole length of the strike.

The Low Level was cut to a good size being 6ft 6ins high and 4ft 6ins wide, and was railed nearly to the forehead. This huge undertaking must have been a great disappointment and the failure of the shaft working, the final straw.

14 Dubhow Mine

Opposite Deepdale valley lies the formidable shoulder of Angle Tarn Pike and here, on either side of Dubhow Gill, is the old Dubhow copper mine. The higher working on the north side of the gill is the older site, of which now little remains apart from a small amount of spoil and a ruined hut. The mineral rights were owned by Edward William Hasell Esq, Lord of the Manor of Dalemain.

The first recorded workings here seem to have been during 1761 and 1763, and these would probably have been at the higher site where a small sump was sunk on the vein. Just above this a small trial was commenced. The copper extracted, however, was of little value and the venture failed.

Much later, on the 18th of October 1850, a small mining company was formed by three men, namely, Jacob Craig of Knock, John Rudd and Jonathon Bainbridge, both of Dufton, all from the county of Westmorland. The conditions of the take-note seem quite harsh for the size of the venture, as can be seen by the following extract: "On striking the copper vein, every tenth bing of copper ore raised will be the royalty paid, the ore to be washed clean and ready for the smelter. No less than two miners are to work the vein and to be employed at the mine. You are required to stone arch and wall all levels and shafts where the same will not stand without support. In all other matters you are to refer to the lease of the Greenside Mining Company. When the vein is struck, you may apply for a full lease of 21 years".

Despite these covenants, work began on the level some 120 feet below the old workings, although the usual distance between workings at this time was sixty feet. There is no apparent documentation by which to ascertain the distance that this company drove the level, but there is one interesting fact which may be considered. At a point some fifty-six yards in from the portal, the level is cut wider and here, resting on a pile of deads was found a pair of clogs. These would often be left by the old men at the end of drive as a gesture of good luck to the next miners, and also as a mark of respect for them. At this point the vein is completely barren and the host rock extremely hard and so, alas, it seems there were no riches to be won for this company.

The next company to take over the mine was set up by John Vipond, a saddler from Penrith, Parker Mandall, also of Penrith and a grocer by trade, and George Little, a miner of Patterdale. They all signed the lease on the 3rd of June 1862 with Edward William Hasell of Dalemain. The contents of the lease differ from the previous one insomuch that the royalty to be paid was to be 1/12th for lead and 1/15th for copper, washed and ready for smelting. Also the company was to ensure that none of its workforce poach for game. Subject to agreement, after one year a full lease could be drawn up for twenty-one years. The mining was pursued with some vigour, no less than four cross-veins being cut, all tried, but copper found only in the last one near the forehead of the main drive, which is some 225 yards in length and is mostly driven on an east-west heading.

At a point on this drive, 162 yards in from the portal, is a crosscut of some significance. It has been driven south for a distance of about eighty yards following a barytes vein of up to six inches in width, and at its widest point a little stoping has been done, the area having been cleared by use of a wheelbarrow. The dimensions of the level here are much smaller, being only 5ft 9ins high and 3ft 9ins wide. The main drive was partially railed out for the use of tubs and consequently is of greater proportions, being 6ft 6ins high and 4ft 6ins wide. The ore taken from the side drive would have amounted to only a few tons and would certainly have contributed little to paying the way for the operation. Again, it is difficult to imagine the tenacity of these men who worked underground. The total distance of the drive is some 337 yards with no ventilation, and conditions at the face would have been of stifling, foul, dust-filled air, a terrible place to have to work for a paltry return.

15 Bleawick Barytes and Deepdale Iron Trials

Under the majestic pinnacle of Place Fell, high in the shattered scree above Bleawick and lying beside a little stream, is to be found a small barytes trial. Driven in the Borrowdale series rock, it follows an east-west barytes vein for a distance of thirty-three yards to a forehead. The actual vein is just over six inches wide and contains some interesting pink barytes. The level has been hand-driven and is of good size which would indicate that it was driven around the 1860s. Numerous quartz veins are to be seen in the vicinity but no further trials. The operation could well have been carried out by the adventurers at Dubhow.

Deepdale is the longest and bleakest of the valleys which cut west towards the spinal ranges of Helvellyn and Fairfield. Underfoot, the head of the valley floor is normally awash with water. Our interest, though, is at the valley mouth just above Deepdale Hall.

The north side of the valley has been cut by two veins, the northernmost called Watson's Vein, presumably after the mining family who lived in Patterdale. It was the Watsons who were credited with informing Captain Borlase of Greenside Mine of yet another vein, this one bearing lead, in the Grisedale valley. However, this vein runs north and south and has been tried by at least two levels. The lowest of these at OD 850 feet is still open and extends sixteen yards to a forehead, and beneath this is a small open-work.

The main level is slightly higher and shows some fine kidney ore and barytes on the dump. The entrance was stoutly constructed with stone walls, and every four feet or so they were recessed to allow heavy timber uprights to be positioned as supports for the horizontal roof timbers. After a few yards the miners made contact with the bedrock, and this revealed the vein lying on the north wall of the level, between a foot and eighteen inches of barytes and hæmatite. The level itself was driven only a few yards but was of very large dimensions, resembling more a large pocket than a level. The extension of this vein can be seen in the small stream just to the south in Low Wood.

This working was operated during 1877 by the Moss Bay Iron Steel Company of Whitehaven, a company which had massive holdings in the West Cumbrian iron ore fields. It is not known why they should have been interested in such an insignificant hæmatite vein, with poor transport facilities and not much potential. The venture was indeed doomed to fail which it did after a short period of prospecting, during which time only a few tons of ore were raised.

Top: Myers Head Mine — remains of waterwheel pit and stone piers which carried the launder to the wheel. Bottom: Hagg Close — the shaft of Hagg Close Mine was situated immediately in front of the farmhouse on the left of the picture. (photos, Ian Tyler)

Myers Head Mine with the waterwheel pit in the foreground and the spoil heaps behind. The shaft was situated beneath the tree on the right of the picture. (photo, Ian Tyler)

John Martin Shaw pushing a tub of ore from a level at Hartsop Hall Mine during the 1930s.

No.2 Level, Hartsop Hall Mine during the 1930s when it was being worked by the Shaw family.

W T Shaw Jnr and Thomas Shaw in Hartsop Hall Mine during the 1930s. Note the hand-held carbide lamps.

Top: the family Shaw — Bill Shaw Jnr, J Shaw, J Martin Shaw, Thomas Shaw and Old Micky the dog at Hartsop Hall Mine. Bottom: the crushing and washing plant at Hartsop Hall Mine.

An 18th century level in Eagle Crag Mine. Note the wooden rails with iron runners and the hydrozincite mineralisation on the walls and tunnel floor. (photo, Ian Tyler)

16 Hagg Close Mine

The first records of this area being worked come to light in a document dated 1753, which refers to Mr Hasell as lord of the manor of Patterdale, Barony of Barton and his ancestors as working the mines without any interruption from any of his tenants, by deed of King Charles II.

Matters were to change however, and in the year 1799 a man called Edmondson obtained a lease to work the mines and veins of mineral in the area. At this time, by chance, servants of John Mounsey of Patterdale Hall, who was the tenant landowner but with no mineral rights, were digging a drainage ditch in one of his fields and exposed a lead vein. Mounsey, a powerful man of means within the area, and interested in mining, went down to have a look at what his men had uncovered. Knowing of the exploits of the miner Edmondson, Mounsey invited him over to inspect the vein and enquired if he would like to work it. Edmondson, a man skilled in his trade but of little cunning, agreed to work the vein. Mounsey, having had his suspicions confirmed by the miner that the vein was indeed worth working, went immediately to see Mr Hasell, the lessor, with whom he was acquainted, and arranged a lease for himself. The enterprise, however, backfired on Mounsey, for after digging a series of pits the vein produced little ore and he was obliged to pay the sum of 15/- as part of the 1/8th royalty imposed by Mr Hasell.

In about 1806 a well known local miner, Mr Little, became interested in the old workings and invited a Mr John Brown from Dryburn near Alston to become mine captain of the new mine he wished to open at Hagg and the Close. They arranged to meet Mr Hasell's steward at the mine site, and Mr Mounsey just happened to be on site and offered to show them the mine and where he had worked the vein previously. Mounsey indicated that a lease could be drawn and that he had no objection to them operating on his land. Mr Little, realising Mounsey's interest in the venture, offered him a partnership, but that gentlemen declined.

A lease for twenty-one years was granted on the 12th of February 1807, the terms of which imposed a royalty of 1/8th on lead and copper raised. This, however, may have been precluded by a take-note of one year as, by that date, work seems to have been already well advanced. By April 26th of that year, Mr Little and miner Alexander Patterson were working forty yards inside the level, which would have taken at least six months to drive. It is also recorded that Mr Mounsey called on numerous occasions to see how work was progressing, even to the extent of, on the 20th of January 1807, offering not only his advice but actually helping the men to dig in the Close.

By the 26th of September 1807 two sets of workings had been tried in the Close and the Hagg. In the Hagg, three pits had been tried on the vein which cuts east to west through from the Close. By December Mr Little was happy with the way the workings were developing and decided to backfill the old pits including the ones Mr Mounsey had worked previously. Their own trial pit had now been cut by a level and a proper shaft working had been put down. The working was now extensive enough to warrant a waterwheel for pumping, windlass for hauling out the dross, and launders to feed the wheel.

Things were looking good and Mr Little, in anticipation of a good year, had engaged six more miners. He knew that ore was standing in the face with more ready to be brought up the shaft, and the first load to be wound up totalled seven bings and four hundredweight (a bing is equal to 8cwt or 407kg). The lead was sold for £15.4s.0d per ton and was taken away by the local carter, Thomas Jackson. Mr Mounsey had supplied much of the timber for the mine, some of it being used for launders and for lining the shaft. The heavy castings for the waterwheel were supplied by the Alston Foundry, duly paid £6.12s.2d.

As the surrounding fields and meadow areas were growing, Mr Mounsey requested that all mining and supplies were brought on site via Grisedale Bridge and not over the track by Hagg Close, and the mining company were happy to comply with this. Up to this point all dealings between the company and Mr Mounsey had been amicable, but things were about to change for the worse, most likely due to the fact that the mine was showing promise and Mounsey was missing out.

Suddenly and without warning, Mr Little was approached by Mounsey who insisted that he had failed to repair ground, and demanding £10 compensation. The next onslaught was a letter dated the 27th of December 1807, which alleged that Mr Little and his partner, John Monkhouse, had been informed that they were trespassing, and that Mr Hasell had no right to allow digging on his land. In fact, Mounsey rented the land from Mr Hasell for 6/8d a year. The letter also inferred that the grant given by King Charles II was invalid with regard to the mineral rights!

Things came to a head when the whole affair was put before a local court. Mr Mounsey's plea reads something like this:

"Between February 11th 1807 and January 16th 1808, John Little and John Monkhouse broke in and damaged the land with their feet and the feet of their servants who did tramp upon unspoiled grass, for mares, geldings, cattle and horses there growing and of great value: £100.0s.0d. Also tore it up with the wheels of carts and wagons covering 10 acres, also with his servants with spades and shovels, mattocks, pickaxes, iron instruments sunk and made 20 holes, pits, shafts and mines. 100 cartloads of earth, stones, sands and gravel, lead ore, timber and as a result John Mounsey was unable to enjoy his closes. Also they dug a level or sough 500 yards by 10 yards wide, 200 feet deep in both closes, erected 4 engines, 4 machines, 4 mills, 4 wheels, 4 pumps, 4 erections, the land is greatly injured and damaged and of little use to him by the 1st March 1808, the chattels and the earth and the 2,000 cartloads removed etc. are worth £20,000."

The outcome of the trial is somewhat of a mystery, but due to the absurdity of the claims and the exaggerated descriptions of the damage, the case was probably either thrown out, or a small damages claim paid and the matter resolved without further ado.

During this time the mine appears to have carried on working despite the trials and tribulations, and by the 1st of May 1808 a further thirteen bings of ore were raised. There were still six miners working underground at the shaft supervised by Peter Patterson, the underground foreman. The men worked six days a week for 3/6d per day and had so far worked fourteen weeks and two days for which they had been paid £15.1s.0d. During this same period one miner, John Carter, was set to driving the levels off the shaft and for this work he was paid 21/- a fathom. He had driven forward seven fathoms, so received a further £17.17s.0d for seventeen weeks work, which made his total pay for the period £25.4s.0d. Other costs accounted for at this time were purchases such as gunpowder at £1.15s.0d, and a copper pricker at 6/-. The mine continued to work at least up until the 1st of December 1808, when eight men were employed at the mine.

Miners who worked for the Patterdale Mining Co 1806-1808

John Little	Director
John Monkhouse	Director
John Brown	Mine captain
Richard Wallace	Miner
John Harrison	Miner
Thomas Harrison	Miner
Joseph Brown	Miner
Richard Stephenson	Miner
Peter Patterson	Miner
Alex Patterson	Miner
Thomas Hill	Miner
John Carter	Miner
Jonathan Patterson	Miner
Thomas Hodgson	Miner
Miles Milburn	Miner and drill sharpener
Thomas Milburn	Ore washer
James Bushfield	Ore washer
Edmund Bulman	Blacksmith
Thomas Eland	Joiner

The mine unfortunately did not prosper for much longer, and no doubt Mr Mounsey reclaimed his land and returned it to pasture once again. Towards the end of the 18th century, Home Farm House was built beside Hagg Beck, fitting idyllically into the surroundings.

In June 1979, one afternoon when the farmer Mike Beaty was working the field next to the Close, his young son Malcolm came running to tell him that a large hole had appeared in front of the house. It certainly was large, for the house had been built virtually on top of the old shaft and crosscut. The local fire brigade started to pump it out but gave up when they discovered the extent of the workings beneath. The local surveyors, however, satisfied that the house was in no immediate danger of collapse, had the hole filled with concrete and rubble. Once again the secrets of Hagg Close Mine were hidden away forever.

17 Hartsop Hall Mine

There are some sources which claim that the Elizabethans may have worked here, but exciting as this notion is, there is absolutely no evidence to support it, either physically on surface and underground or in any documentation. Given the fact that the records of the German mining operations were meticulous in the extreme, it would seem unlikely that any working here would have gone unmentioned. Certainly the mine has a long history, dating back to the old men who toiled hard to eke out a living and hack out the back of the vein where the mineral was exposed before the use of gunpowder for blasting.

The direction of the vein runs NE-SW in the Borrowdale volcanic ashes and the main minerals taken were lead and barytes. The vein varies in width from a stringer to up to three feet, the matrix is quartz and the galena yielded up to thirty ounces of silver per ton, a much higher content than the Greenside Mine assay. The barytes, which is plentiful, has a pinkish tinge, and lesser minerals to be found are blende, straw-coloured fluorspar, wulfenite and copper. It has also been reported that a little gold is to be found in the galena but certainly this is of no commercial value.

The mine is to be found hidden in the dense woodland which abounds on the southern slopes of Hartsop above How through which the vein passes. The old manor house called Hartsop Hall, from which the mine takes its name, is just below the trees and has seen the many changes which have taken place in the valley since the earliest habitation

At the side of the track from Hartsop Hall — heading in a south-westerly direction towards Hogget Gill — are the remains of an old stone circle, sure proof of earlier settlers. About half a mile from here lies one of the oldest remaining smelters in the Lake District. Built beneath the savage ravine of Hogget Gill, here the ancients used the pounding, rushing water to wash the ore and to power a small waterwheel which drove the bellows for the hearth. The smelter was housed in a small cobblestone building measuring 24 feet by 16 feet, and divided into two. One half contained the actual furnace, which was of the open hearth type, the fire heat being intensified by the bellows driven by the overshot waterwheel. The remains of the supports for the launders and the groove which brought the water can be traced back to where the stream had been dammed to ensure a constant supply. The other half of the building was used to collect and store the smelted ore.

This small type of operation would have been worked by two men, a smelter and a labourer. The fire was of charcoal and once the correct temperature was achieved, would have been capable of smelting a ton of ore at a time. This method was very basic and an analysis of the actual residue slags from the site reveals a figure of 12% to 18% lead remaining in the slags. Also very much in evidence in the residue is barytes, which would indicate that the ore smelted here was from Hartsop Hall. The fact that the smelter was fired by charcoal also precludes the Elizabethan theory, as wood was wholly in use at that time for smelting, and the more sophisticated system of using charcoal as a smelting fuel did not begin until around 1700.

The first lease for the area is dated the 17th of April 1696, and is between John Lowther of Whitehaven on the one hand and Nathaniel Johnson, an alderman from Newcastle upon Tyne, and Robert Jopling, a country gent, who were granted the lease to work Hartsop Hall vein for twenty-one years with a royalty payable of every ninth bing of lead. It would be reasonable to assume that this was the first occupancy and that they successfully worked the vein from the surface by a series of bellpits and open works. They could also have been the adventurers who tried the large exposed vein near to the site of the smelter, if this contained lead. Indeed, this may have been really the only reason for selecting this remote spot as it is so far from the main Hartsop working. Whoever it was who built this place certainly went to a great deal of effort, and a good road capable of taking a large cart has been engineered from the hall right up to the smelter site. It is interesting to note that the date of this first lease would also coincide with the type of smelter built.

How long the previous tenants remained or what the totals of ore raised were is not certain, but they most certainly left their mark for others to see later in this remote place. The next occupants appeared to be much more flamboyant, or certainly the company's name was: The Governor and Company for Smelting down Lead with Pit Coal and Sea Coal, Saint Martin's Lane, Cannon Street, London. This company, which later became known as the London Lead Company, had extensive mining operations in the Alston area. Mr Thomas Dodd, their chief mining agent, negotiated a lease dated the 25th of November 1802 with Lord Viscount Lowther of Lowther Hall, to work the vein at Hartsop Hall. The term was for twenty-one years at a royalty of 1/5th of clean, washed ore produced. The venture was short-lived however, some ore being raised, and when the company departed they left a parcel of washed lead ore weighing 10 cwt, being the Lord Lowther's royalty.

Assuming that the company were honest operators, this would mean that the total amount of lead raised would have been 2 tons 10 cwt. In a letter dated the 7th of July 1804, Thomas Dodd informed Lord Lowther of his departure from the site, assuring him that all workings and shafts were secured, and also inferring that there was no future here. It is most probable that the site was vacated before this date and that this was the final winding up of affairs. It is surprising that such an important company did not show any interest in Greenside.

Another twenty-one year lease was granted by Lord Lowther to a Mr R Tipple, also of the "Governor and Company" on the 12th of September 1803, at a royalty of 1/5th. During the period of this lease a total of 6 tons 5 cwt of lead ore was raised and the total expenditure during the working was £283.9s.2d. The mine would have worked for about four years during this time, and the workings would have still been hardly developed. It is possible that it was this company who cut the No.1 Level which is a crosscut intersecting the vein twelve yards inside. The level is small being only 5ft 9ins high and 2ft 6ins wide, and at the entrance a small hand-dressing floor is still to be found.

There seems to have been no activity then until, on the 17th of August 1830, a take-note for one year was granted to Mr John Slack and Mr E Proud, two experienced miners from Alston. The royalty had now altered to 1/6th, possibly an attempt by the lord of the manor to make the terms more attractive as the the cost of drawing up leases must have been considerable. The fact that the new lessors were from Alston and were perhaps already known to Mr Tipple had prompted him to appraise them of the mining prospects here, and that the mine could be ideal for working by a small company. The Governor and Company was now well in control of the Quaker Company, generally known as the London Lead Company, and considered this site be to too small in comparison with their operations over in Alston and Nenthead.

The Hartsop venture prospered for no more than two years. The two Alston companies were very experienced and almost certainly would have driven No.1 Level at OD 900 feet and the No.3 Level at OD 777 feet, which cuts directly onto the vein. Adjacent to the entrance of No.3 are the remains of a substantial hand-dressing floor. From here the old miners' path meanders up beside the vein past the old pits and open works to No.1 Level.

A further lease was taken up by a Mr Bowman on the 12th of June 1832 but this appears to have been a short-lived venture.

Why nothing was done at Hartsop Hall for the next thirty years or so is a mystery, but perhaps all attention was being devoted by the prospecting companies to Greenside down the road. Certainly the population in the valley had grown and work was plentiful both underground and on the surface for the local miners and tradesmen alike.

Eventually, in 1863, a company was formed and a lease taken up. The directors were a Mr Kennedy, Mr Whiteside, Mr John Jameson of Brougham in Westmorland, and Mr George Head-Head of Rickerby House in Carlisle, who was a local banker. Both Head-Head and Jameson were directors of the Greenside Mining Co. The full details of the lease are unknown, but a dead rent of £10 a year was payable on top of the royalty.

This company, with its financial backing and influence through George Head-Head, would have ensured the initial development costs of the mine. The plans were to first drive a new crosscut forty-eight feet below the old No.1 Level. The crosscut would be known as No.2 Level and on cutting the vein work would commence on driving both north and south to undercut and drain the old men's workings. By the entrance of the No.2 Level would be built a small sorting area, and from there would be constructed a substantial incline about a hundred yards long and ten feet wide, so that ore could be transported by cart to the bottom where a mill was to be built. Most important perhaps was the plan to drive another crosscut level to the vein near the mill site, which would be known as the No.4 Level. Eventually, this would undercut the whole of the existing and planned workings, and all the newly won ore would come straight out to the mill. Such was the plan, and certainly a very great deal of it was accomplished.

In the new mill it was intended to install a steam engine for the crushing of the ore, however the cost of coal proved prohibitive due to the long distance of cartage. Instead, a substantial water leat was constructed to bring water from Hartsop Beck into a large reservoir which was sited a few hundred yards from the mine. This ensured a steady flow of water to a waterwheel which in turn drove the crushing machinery. Some of the excellent stone water leats are still in evidence and are a well preserved credit to the miners who built them.

Just above the No.4 Level a second stream was dammed and the water diverted to join the leat. The problem of power having been solved, all energies were now turned to the working of the mine under the ever watchful eye of the mine manager, a Mr Phillips. The driving of the levels, however, was marred by a frightful accident. In 1864, whilst setting a charge, miner George Crosthwaite from Troutbeck was blinded and severely injured by the blast. Although he survived the accident he was was reduced to the pitiful act of begging on the streets of Penrith for the rest of his life.

The company, in an effort to prove the northern extent of the vein, drove a short crosscut at OD 900 feet, just north of No.1 Level. Now known as Wood Level, it does in fact cut a lead vein at the face twelve yards in. Nearby, in the thickly-wooded hillside, is to be found a small quarry.

Work continued steadily, and by 1865 the No.4 Level was well developed and a sump had been put down in the sole by a miner called Heslep, assisted by miner William Hellen from Low Hartsop village. In the roof, the stope had cut through into the No.3 Level which was the old men's lowest working, and from which a great deal of ground had also been stoped. They had built a fair sized stone hut near to the entrance of the level for washing and sorting the ore, the ruins of which can still be seen. For the new company, this area had also proved profitable and some good ore was taken, whilst in the No.2 the vein was being worked north and south in good ore. The miners had also cut through to the old workings above, and this ventilated the mine considerably. Ore was carted down to the mill area after being sorted outside No.2 Level, and this ensured that the carts only ferried good stuff to the mill.

In March 1867 two miners driving north in the face of No.4 Level were paid 70/- a fathom, whereas three other miners who were working north of the stope and rise up to the old No.3 Level were paid only 35/- a fathom, so this ground must have been much softer. It was near to this point where miner William Hellen put a sump down in good ore.

Sometime later, in 1867, there was a good deal of unrest at the mine, with the men complaining about poor wages and their treatment at the hands of Mr Phillips who turned out to be, by all accounts, a ruthless taskmaster. His main concern was whether or not the mine was producing sufficiently and at this time it most definitely was, with ore being mined from various areas. In No.2, the level south was directly above Hellen's Sump and some small pockets of ore had been stoped on the way. Now, the men were in the process of putting a sump down to No.4 which was to be known as the Ore Pass Sump, and to assist the speed of its completion men were also rising from No.4. This would eventually connect the two levels thus allowing all the ore to be passed down the hopper and trammed directly to the mill. A manway was also being constructed for miners to pass between the levels. Production was between twenty-three and twenty-four tons of lead a year, with development work going ahead and ore there to be got.

By the 7th of April 1868 there were nine miners working the mine. In No.2 Level, two sumps had been sunk in the sole of the northerly drive, the first one known as Joe's Sump, above which a considerable amount of ore had been previously stoped. From this point the vein had nipped in and a barren drive of eighty yards had been driven before coming back into good orey ground. Here a further sump was put down fifteen feet to test the vein, and this was known as Jon's Sump. Above this the vein was stoped out into the roof to a height of ninety feet. The ore here was good and the area was worked vigorously.

The extra material produced brought the mill up to full operational capacity. It seems as though this extra material must have caused a certain amount of mill effluent to spill into Hartsop Beck, and the local tenant farmer, Mr Backhouse of Hartsop Hall, could not fail to notice the discoloration in the water. He was quick to start making serious allegations that the mill tailings were killing his livestock and ruining his land, and also killing the fish in Brothers Water.

Work at the mine progressed steadily and ore was brought out on a regular basis, but Mr Backhouse became a thorough nuisance and, living so close to the mine, was a regular, albeit unwelcome, visitor. Matters came to a head on the 11th of November 1868 when Mr Backhouse engaged a lawyer, Mr Benn, and the company sought the services of Mr William Heskett of Plumpton Hall, near Penrith, to act on their behalf in a bid to resolve the matter by arbitration.

Winter passed, as did the summer of '69, and still the problem of the effluent remained. The mine owners had spent some considerable time on trying to rectify the matter, effecting various means to alleviate the discharge, but to little avail.

Promises were made by the directors that the rectification work would continue whilst the business of mining ore was proceeding, but Backhouse was an unremitting character and continued to complain and harass the miners. As a result of all this the production figures for the year ending the 18th of November 1870 were the lowest to date with virtually no ore having been raised. Royalties were not paid for the year, and the only money received by the lord of the manor was the £10 rent. This, naturally, was not good enough.

As the mining company had still failed to achieve either agreement or compromise with Mr Backhouse, and after much discussion with Mr George Lumb, the Earl of Lonsdale's mining agent, the company was given notice to quit on the 14th of November 1871.

For some years the mine remained closed and gradually became derelict. Many of the buildings, such as the mine office and smithy, were pulled down, and some of the stones and slates used to build a new barn at Hartsop Hall Farm. Much of the material was also used to extend the Brothers Water Hotel. However, the small store complete with stone chimney is still to be found just up from the mill below No.4 Level. The waterwheel was sold to the Horrax bobbin mill at Ambleside, which supplied bobbins to the Lancashire cotton industry, and the rest of the equipment was sold piecemeal.

It is a sad fact that the demise of the mine and the subsequent loss of men's livelihoods was brought about by a belligerent farmer whose claims were, it appears after all, unfounded. Tests had been done on various effluent waste samples of mine water being discharged into the streams from the mills both at Greenside and Helvellyn mines. A Dr Pole, working for the Royal Commission and Water Supply stated: "the char had reduced in quantity, but both here and in Brothers Water, the trout were fine and healthy". It was his considered opinion that if the water were to pass through settling tanks it would be fine for human consumption.

Further tests made by a Dr Franklin revealed only three cases of contamination, at Glenridding and Thirlmere, but this was so minute in quantity as to allay any apprehension of injurious consequences. Indeed, this slight contamination actually had the effect of improving the quality of the water, by helping to purify the heavy peatiness it contained from its source in the fells. Dr Pole added that with regard to the problem of Mr Backhouse's livestock and fish, it was possible that they could have ingested quartz particles from the water, though no lead existed.

During August 1888 the Greenside Mining Company sent a group of miners to Hartsop Hall, headed by Mr John Comer the underground manager, to assess the mine. If ore was found no milling would be done here in view of the previous problems, and only good ore would be carted to the Greenside mill just down the road. No doubt it was the high silver content in the lead which created the interest.

The group of men walked past the ruined smithy and mill remains, nearly eighteen years after the mine's closure, and the place was overgrown with vegetation. The men started up the incline, ducking their heads to avoid the branches of the trees and shrubs. On arriving at the portal of No.2 Level, Mr Comer sent a couple of men off to inspect the No.1, and the rest of the men lit their candles and prepared to follow the boss. The crosscut was clear to the vein and he decided to follow the level south. On seeing the lights, the men above called down, advising Mr Comer that the No.1 was down from surface. The No.2 Level continued for forty-four fathoms until the Ore Pass Sump was reached and the ladderway down through the stope to No.4. Nearby was a rise going up to a sub-level and one of the men climbed up the thirty or so feet, using the footholes which were cut in the rock and an old chain for support, his lantern hanging on his belt. His mates below tried to assist his progress by holding up their lamps and shouting words of encouragement. A slip here would mean a fatal fall 150 feet down through the ore-pass onto the No.4. Our man completed his climb safely and shone his lantern into the old men's workings, alas only to find it completely run in. Precariously, he reversed the climb and was thanked by John Comer for a brave effort. They then returned to the crosscut to be met by the men who had, by now, descended from the No.1.

Comer noted the extent of the stope at this point and as they continued they came, almost immediately, to Joe's Sump, which was flooded to the collar. One man set to to plumb this while the others went on. The level was standing well, considering that it was so near to the surface, and Jon's Sump was reached and also plumbed. By now John Comer could see the massive stope before him, and two or three of the men went ahead, lights twinkling in the darkness. One miner commenced the climb into the stope, up a fixed ladder hanging on the footwall. Up and up he went into the blackness, his light growing fainter and fainter. On reaching the top he shouted down that all the ore had been taken, and then, his task accomplished, slowly descended the ladder back to his mates and comparative safety. The level was a total of sixty-three fathoms in, with just a small stringer of lead in the face. The men plumbing the sumps reported that Joe's was down ten fathoms and Jon's down about three fathoms.

It was apparent that there was very little workable ore value left in these two levels, so the group of men made their way out to day and a spot of bait. As they sat at the level mouth, no doubt some of them were recalling how their fathers had worked this mine, and what a nuisance old Backhouse had been. Bait time over, Mr Comer asked a couple of men to go through to No.4 Level by the Ore Pass Sump. The rest would cut down the incline and into the No.4 adit and they would meet up later. As the group proceeded down the crosscut, their voices became muffled and the air smelt stale. Some of the men started to hang back, aware of how quick an effect bad air could have. There was not very far to go however, as the level had collapsed, and a quick look showed that the deads from the stope above, and possibly the old men's working, were in on top. Shouting and hammering to establish contact with the other two men was to no avail, and the physical exertions had made the air now really foul. Mr Comer ordered his men to vacate the spot and needless to say no second request was needed. Shortly afterwards, the other two miners arrived and reported straight away to Mr Comer. Although the ore pass was stable, the stopes were in a loose condition, and some old rails were still in situ, these being the wooden type with metal strips on top. As the men followed the level back out to day they passed through an area of barren ground and came across a stope of extensive proportions, and a few yards further on they came to backed-up water and a large roof-fall. The day's work now over the men, after a word of thanks from the boss, started their walk home. As he passed them in his gig, Mr Comer looked thoughtful. The No.4 Level had been a disappointment, a distance of only sixty-four fathoms being directly accessible, but after nearly twenty years of neglect it was only to be expected. The report was duly presented to the directors of Greenside and, as we now know, the option on the lease was never taken up.

There must have been many times when Thomas Shaw looked thoughtfully from his slate workings on Caudale Moor at the now long derelict lead mine of Hartsop Hall. Thomas Shaw was a well known and respected miner in the area, his family history steeped in mining tradition. Over the years he had talked with many an old miner and listened to the tales of days gone by. Now he recalled that one old miner from Coniston had told him of a good pocket of ore he had seen, years before, in a sump he was putting down in Hartsop Mine, and that this, to the best of his knowledge, had never all been taken. The miner's name was William Hellen.

Perhaps now was the time to fulfil a dream of working one of the oldest mines in the valley and proving the old vein. A cursory inspection both on surface and underground left little to the imagination as to what sixty years of total neglect had done, and many men would not have bothered any further. However, our Thomas was a man of determination and after lengthy discussion with his family a company to be known as the Hartsop Hall Syndicate was formed. The workforce consisted of himself, his brother-in-law Jonathan Myles, and his son W T (Bill) Shaw. A lease was drawn up and work commenced on the 28th of March 1931.

The first job was to clear No.2 and No.4 levels, and iron rails were laid to the fall in No.4 to start the work of clearing the dross. Conditions were better now for lighting was by carbide lamp, which would burn for up to ten hours at a time and gave a substantial light. After much hard work and the aid of a pony pulling out the waste, the big fall was cleared, but it must be assumed that Heslep's Sump was not worth the time as no heed seems to have been taken of it. The far side of the fall revealed the old mine just as it was when the last men had worked there. The elm rails with metal strips and the oak sleepers pegged with oak were still down, and as Thomas and his partners walked along the level they saw the marks of the old men's clogs crisscrossing in the mud. Here and there patches of blackened soot shadowed the walls where the tallow candles had been, such a poor light compared to the carbide lamps.

The air was freshening now and smelt good, and they had arrived at the Ore Pass Sump from the No.2 Level, 150 feet above. The lamps flickered in the draught, and just ahead they saw the collar of Hellen's Sump, but this, to their dismay, had been back-filled. To compound this setback they also saw that, on looking upwards into the stope, much of the area above was in a very unstable state and was in danger of collapse. It seemed certain that no work could be pursued here. The day had been one to remember and no doubt Thomas Shaw gave much thought to the project as he and his family walked back to day. An inspection in the No.2 Level showed that nothing had changed since Comer and his men had viewed the site in 1888, but there were a few pockets of ore to be had, and the sumps would have to be pumped out to reveal what the ore was at depth.

Work began both underground and on surface and before long the mine was operational. A small mill was erected on the old site consisting of crushing rollers, two jigs and a Diester table, all belt-driven by a 16/20 horsepower Marshall oil engine. This also worked a compressor for the single air rock drill. Mechanisation had finally arrived! A local miner, Bill Hebson, originally from Threlkeld, now joined the team and was paid between 5/- and 7/- a day. He was, to his credit, reputed to have been one of the hardest working and most conscientious men in the dales.

It was decided that the No.4 Level would be tried, and at Hellen's Sump they started to remove the capping. The old men had back-filled the shaft set on timber, and as the water was being pumped out the whole lot collapsed downwards into the bottom of the sump. The men could see below them the old jackroll and kibbles, but these were carried to the bottom of the sump and lost forever. During this operation, the water which had been stagnant for some seventy years, and had stood against the rotting timber and old ironwork, gave off a quantity of sulphuretted hydrogen. One miner was overcome and another affected by the fumes, and both were lucky to recover. Had this happened at a blind heading and not as it did, near to the Ore Pass rise, the outcome might indeed have been a tragic one. No further pursuance of the sump was made, and then, the price of lead having fallen, the whole project was halted temporarily.

Just two years later, however, Thomas and his men were back with renewed heart to recommence where they had left off. The mill was extended and a crusher, a three-head stamp battery, and a slime table were installed. Problems were experienced with the separation process as the galena was very heavily mixed with the barytes and recovery was only about 60%. The mill was installed in prefabricated buildings of either tin sheet or wooden planking.

Work commenced in the No.2 Level and Joe's Sump was cleared and good ore won. Compressed-air lines were installed and the miners started to draw the ore in the face of the north stope. The vein was followed in a pilot incline drive known as the "Birdcage" where the lead vein is between two and three inches of solid galena. The work continued slowly and the level was stoped out directly under the old Wood Crosscut, some ninety feet above.

Around the outbreak of war the workforce was joined by two young men from Patterdale, Billy "Bunty" Smith, whose father was a blacksmith at Greenside, and Edward "Teddy" Flitters, son of the landlord of the Brothers Water Hotel. Major changes were afoot, however, when in 1940 the company was joined by a Mr J H Clay with a capital injection of £2,000, and the Hartsop Mines Ltd was formed.

A large 85 horsepower Crossley diesel engine was transported to the mine by a local haulage company, and it was with difficulty that it negotiated the lake road beside Brothers Water. Also installed, to increase drilling capacity, was a compressor plus two more tables for better recovery of ore. Bill Hebson was now underground foreman and had four or five men working underground or on surface. Wages had risen to 10/- a day.

Work was continuing at a steady rate and No.4 Level was a hive of activity. A level was pushed on to bypass the unstable ground of Hellen's Sump. All the driving was done by compressed-air drills, of which they now had two. The drilling was all done dry, and this created so much dust at the face that it was almost impossible to see, and the men, on returning to day, were caked from head to foot. Charges were laid and set off by lighting safety fuses, the shot pattern being normally twelve holes at 24 to 36 inches deep. After firing, all the mucking out was done by hand, the waste being trammed out to day apparently also by hand, the horse having now gone.

These were happy days and the war seemed far away, although Teddy and Bunty were soon to go, as was Bill Hebson, but not before they had tried Tom Shaw's personal artillery, an old blunderbuss which hung in the office. On a warm summer's day, while Tom Shaw was away on business, the men were sitting around having their bait and the conversation got round to the gun. It was too good a chance to miss, and Bunty went to get it. John Shaw remembered he had just taken an engine to pieces and had loads of ballbearings, and Teddy was picking up nails and all manner of rubbish to ram down the trumpet-shaped barrel. Bill went off to get some powder from the store and, the gun charged and loaded, a suitable target was spotted — a poor old crow resting in a nearby tree.

As the other three looked on, Bunty aimed the piece towards the tree and pulled back the trigger. The air was filled with a huge, resounding bang, the barrel belched flame and Bunty was flung backwards head over heels with the recoil. The result of the adventure was one decimated tree and a demised crow, and all agreed that for safety's sake the exercise would not be repeated. Happily, these men were to survive the war and return to their homes and their work in the dale.

The mine continued to produce ore, and lead was now fetching around £24 per ton. It was transported by Charlton's Ltd, of Hexham, to Walker Parker Ltd at the Elswick Lead Works, Newcastle upon Tyne, for refining and smelting.

The death of Lord Lonsdale in 1944 brought problems for the company insomuch that the land was sold to the National Trust, who then imposed numerous restrictions on the mining operations. The Greenside company was approached to see if they would accept the ore and process it through their more sophisticated flotation plant, but for some reason they would not oblige. It may possibly have been because of the difference of the ore grade, or the fact that at this time they were already working twenty-four hours a day. Whatever the reasons, this negative decision, combined with the National Trust's impositions, forced the Hartsop company's closure.

Many years later, when Bill Shaw was working for McKechnies of Widnes, he sent a sample of the Hartsop ore to their laboratory for assay and the result was as follows:

Lead	20.36
Zinc	3.80
Barytes	4.40
Silver	5.53
Gold	0.016

No doubt as he parcelled the ore his mind wandered back to the happy times he had had, of all the hardships as well as the adventures. Maybe he recalled the day of the great storm which came sweeping down the gill, which runs past No.3 Level, and exposed a vein of barytes hitherto undiscovered. The vein was tried by his father for a few yards, and it was given the name Thomas Vein, but this was all long ago, and those days were over.

During 1947 a lease was drawn up with Lowther Estates to continue the working of the mine, the plan being to drive a deep level from OD 600 feet and take all the ore at depth through the sole of No.4 Level. The project, however, never got under way, the lease lapsed, and the valley had but one remaining mine — Greenside.

Mill Flow Sheet for Hartsop Hall Mine 1931-1934

5 Ton hopper
Jaw crusher 14 x 4
1 Three-head stamp battery
3 Screens
2 Two compartment jigs
1 Diester sand table
1 Slime table.
Power 16/20 hp Marshal engine.

Mill Flow Sheet for Hartsop Hall Mine 1940

5 Ton hopper
Jaw crusher 14 x 4
Huntington roller mill
2 Deck screens
1 Three compartment jig
2 Slime tables
1 Diester sand table.
Power 85 hp Crossley diesel engine.
250 cu ft compressor.

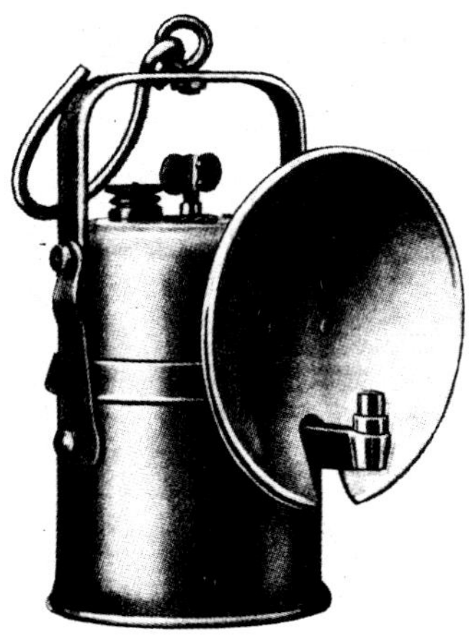

Yorkshire Lamp Fig. 60

MINING LAMP

fitted with

QUICK ACTION ECCENTRIC LOCK

In Steel (tinned after manufacture) with Brass Water Vessel Base

No. 1 size is also supplied in Brass

Three sizes	No. 0	No. 1	No. 2
Height to Bridle ...	$9\frac{3}{4}$ in.	$8\frac{1}{2}$ in.	$7\frac{1}{2}$ in.
Weight	3 lb.	2 lb. 6 oz.	1 lb. 12 oz.
Charge of Carbide ...	16 oz.	10 oz.	6 oz.
Lighting Power	25 c.p.	20 c.p.	16 c.p.
Burning Capacity ...	10–12 hrs.	10–12 hrs.	8 hrs.
Burner No. 105	21 litre	14 litre	$10\frac{1}{2}$ litre
Code Word	CABA	CADDA	CAVA

18 Ruthwaite Mine

As walkers wend their weary way from Patterdale up the old packhorse road to Grisedale Tarn and on to Grasmere, they probably never realise that miners from Grasmere used to walk this way to work at Greenside Mine, a journey of between six and seven miles. Originally they did this six days a week, then, as one Grasmere miner said, "conditions improved, now we only work a five-and-a-half day week". To be honest, the logic baffled me!

The mine at Ruthwaite was sited in an area once called Grisedale Forest, which would indicate that the place had been extensively wooded and would have looked much different apart, that is, from the unchanging majesty and grandeur of the surrounding mountains. The Ruthwaite mines are split into three separate workings, the original ones being just behind what has become known to many as the Ruthwaite climbing hut. This was possibly the old miners' office and smithy which has been upgraded over the years; and in front of this a few small surface trials have been made on a vein outcrop. The other group of workings is to be found 250 feet below, near where Ruthwaite Beck joins Grisedale Beck.

Just behind the hut it looks as though an attempt was made to hush and expose the vein, as there are three distinct grooves into which water from the stream above has been channelled. This proved successful, as the vein was exposed and has been worked down from the surface to a depth of ten feet and a width of four feet. These workings are now flooded, but some lead or copper ore has been taken. Just below the open work, two small trial levels have been driven, no doubt to undercut the open work. The most southerly level extends just thirty feet and shows galena in the face, and a few yards to the north and just above Ruthwaite Gill, is the second level which has been driven twenty feet. Both these levels are of very small dimensions, being less than three feet wide, and both have been hand-drilled. A further small trial or pit has been worked on a lead vein in the gill, just above the open work.

A second vein containing barytes and quartz courses NE-SW and has had some small surface trials on it in front of the hut. At the junction of the two becks can be found the other workings consisting of a single level, which was driven about 120 yards and, in anticipation, a small dam constructed above for water for ore washing. Above and to the north in the next gill are two grassed-over trials. All these workings are attributed to the first three lease holders, the earliest of these being two miners who had a tacit agreement in 1753 with Mr Mounsey of Patterdale Hall.

The first actual lease was issued on the 24th of May 1784 for twenty-one years by Mr William Hasell, and was with John Edmondson and John Walker, both of Keswick, and a Mr Dyer and Mr Walker. A further lease was granted to a Mr Thompson in 1807. Alas, from the information available, it is not possible to be more accurate than this.

A further lease was taken up on the 24th of July 1861 with James Phillips and Thomas Redfern, both of Patterdale, the term being for three years at a royalty of 1/12th. Due to the amount of work already done by the old men, it was decided to try the vein at depth, and this they did by driving the level from Ruthwaite Beck where the large quartz vein is in evidence. Because of the remoteness of the area, a large stone smithy and bothy was built (the Ruthwaite hut), and must have afforded much warmth and comfort to its occupants all those years ago.

The vein has been tried at both sides of the stream, and it is obvious that the main drive is to the south. The level is cut to a good size and is certainly big enough to have been railed out. That this was never done may well have been because they wanted proof of the vein before that amount of commitment was made and, consequently, all material was wheelbarrowed out. To cut a level this size would indicate great expectations, but after the level was hand-drilled and blasted forward for some hundred yards it suddenly became much reduced in size, and a lot of waste material has been left.

It must have been apparent to the miners that the venture was failing as the level runs just past the workings above, where the ore had been plentiful. Here, there is no sign of a mineralised vein, in fact there is only a two-inch copper vein to be seen, about forty yards in. What an awful disappointment this must have been to those men who toiled away in this lonely outpost. When walking out to day from the trial, the thundering of the cascade at the entrance reverberates eerily round the tunnel walls, quickening heart and steps as daylight comes in sight.

19 Brown Cove Mine

The two adits to the mine have the distinction of being some of the highest workings in the Lake District, the uppermost level being at OD 2,400 feet. Initially the level has been cut through solid rock following a lead vein. This could be the East West Cross-course or the Brown Cove Vein which was mined by Captain John Muse at Wythburn on the western slopes of Helvellyn. The good ground soon gives way to a clayey and very soft area where the level is timbered. The lead vein is lost and the tunnel ends in a doukey forehead. The level was hand-driven and the remains of powder kegs are still strewn around. An odd stub of tallow candle remains stuck on the wall in its clay holder. Interestingly, early graffiti of the initials J M and the date of 1802 is a genuine indication of the mine's working date. The initials could be those of J Milburn, J Monkhouse, or J Martin, all of whom were active miners in the area at the time.

Below, at OD 2,200 feet, is an impressive stone-arched portal with the remains of a miners' hut. The level, which is up to 250 yards in length, has collapsed after a short distance. The collapse occurs on striking similar ground to that in the higher level, the timber roof supports having given way. The wooden rails with metal strip tops still lie in situ, but even more amazing are the forty-one rail chairs to be found just outside the portal (these were uncovered by the writer in 1980 and more recently a further twenty have been discovered).

Some 200 feet below this level are the remains of a substantial smithy and office. Water was fed to the smithy by a small leat, and from here a steep zigzag path winds up to the workings. The smithy no doubt gave the men protection from the harsh weather in this isolated spot. On a fine day, travelling light, it is a good ninety minutes walk from the village and all up hill. The top level shows a fair lead vein with barytes, and it was probably because of this that they decided to cut a deeper level in the hope that the vein would improve at depth. This appears not to have been the case, as the solid nature of the rock is not borne out underground. However, much effort and skill has gone into the arching of the level and the building of the smithy.

The use of elm rails would suggest that these lower workings date from around the 1835 period. This early work failed but the Greenside company, in about 1870, intended to develop the workings and extend the lower level, leaving the old elm rails and laying iron rails further in. Tubs were certainly used here; and a great deal of work must have been involved in transporting the heavy gear.

No underground connection was made with the upper working, but this is hardly surprising, the vertical height difference being some 200 feet which is far in excess of normal drift working techniques. There was certainly no revenue from this adventure, the lower tip being virtually barren. On careful examination of the site, to the north-west above the top working, lead and barytes can be found. Also a small wedge was found here by the writer in 1991, and it is possible a further trial was made but is now covered by masses of scree.

20 Eagle Crag Mine

Eagle Crag lies just over halfway up the Grisedale valley, a distance of just two-and-a-half miles from Patterdale village. The mine takes its name from a place where once the eagle soared high above the rocky towers of the crag to glimpse Striding Edge on Helvellyn. This is one of the most rugged and serene areas in the Lakes. In years gone by the valley was well wooded and was a popular area for hunting deer and game, though these were the sports and pastimes of the well to do. For many men, however, the steep, gaunt crag was a place for work and survival.

The east face of the crag is dissected by the main mineral vein, running east-west, which contains copper, zinc and lead, the lead providing some sixteen ounces of silver per ton, and is thought to reach Thirlmere in the west. Eastwards, the vein is to be seen clearly cutting Grisedale Beck after which it forces its course through the shoulder of St Sunday Crag and eventually emerges in Patterdale behind Beckstones Farm. Here the vein shows lead yet it has never been tried at this point.

Another vein, known as the Clay Vein, and containing galena, courses NE-SW and can be seen cutting across Nethermost Beck. The beck, being one of the main watersheds of the high mountains, is difficult to ford during inclement weather, and some of the workings require some scrambling to visit. Dominating all is the massive height of Helvellyn bringing sometimes rapid weather changes.

Eagle Crag Mine is a time capsule having lain undisturbed for many years. Its inner depths still remain untouched, holding fast its secrets, and many of the old buildings are to be found still standing, although some are in urgent need of repair. The workings and the mine buildings are scattered over about two miles from between OD 750 feet up to OD 1,820 feet.

The veins have been tried in many places and the mine is of some antiquity. The first lease was issued on the 24th of May 1784 by William Hasell, of Dalemain in Cumberland, lord of the manor, to John Walker, a shopkeeper and John Edmondson, both of Keswick, and a Mr Walker and a Mr Dyer. The lease was granted for a period of twenty-one years. These gentlemen also worked the Ruthwaite mine during the same period. A subsequent lease was granted by Edward Hasell to John Little of Rays, in the parish of Alston, and John Monkhouse of Penrith on the 12th of February 1807.

These early adventurers were content to excavate the east-west vein where it was exposed to the surface, in much the same manner as the old men had done at Hartsop Hall. The workings were opened up near the top of the hill at OD 1,780 feet, a small hand-dressing floor was set up, and a smithy and bothy built. This excellent little building is still standing and has survived the ravages of more than two-hundred winters.

As the open work became deeper a stone-lined shaft was put down at the end of the working and a jackroll erected to raise the lead ore. The work progressed well and a small crosscut level, the No.7 Level, was driven some forty-five feet below the open work. The level is of very small dimensions, being 5ft 9ins high by 3 feet wide, and it meanders the forty yards to the vein where, on its completion, it successfully de-watered the workings. The lead here was good, and a further sump was put down directly above this point, to a depth of 15 feet, and a level taken east for a few yards. This is known as the No.6, but it did not connect with the workings below. The east-west vein was tried further west by at least two more pits and another crosscut, the No.8, but these were failures.

By now a substantial pony track had been developed on the far side of Nethermost Beck and a small bridge put across for access to the mine. This track can be followed all the way down the valley, where the miners could shortcut over Keldas to Glenridding. If going to Patterdale, they could join the already well established pony route on the southern side of the valley. Near this junction, hidden amongst the moraine, lie the ruins of an old building which, perhaps, housed the miners and workmen on the initial construction of the mine in its early years.

Between the twin buttresses of Eagle Crag the vein splits in two forming a huge "Y" in the face. In the top of the left section, which is the east-west vein, a small trial has been driven twenty yards on a copper string, the dimensions being only 5ft 9ins high by 3 feet wide. The right section has been worked out at surface and some lead taken. Hidden in the top of this vein could be the No.4 Level. Both veins at this point have been cleared of all visible mineral right down the face.

A rich pocket of ore was worked at OD 960 feet and a great deal of lead and copper won. In fact, the ore was worked in some twenty yards and stoped right through to the surface.

In my opinion, this is as far as the old men got with the workings, for tallow candles still remain on the walls, and the grooves left by the work of hand-boring for blasting are still to be seen, thus ruling out any pre-gunpowder operation.

It is difficult today to imagine just how hard was the life style of these early miners. At this time there were no hostel facilities at the mine, and the walk to work was a daily route of some three-and-a-half miles at least, the final mile up very steep fell. The climactic conditions in winter hardly bear thinking about at this altitude, and the snow, ice and wind would prevent any washing of ore. It is possible, in fact, that the mine would be shut down for much of the winter season.

In summer also, one of the main problems would be a shortage of water for ore-dressing, as the water fails to be trapped in Nethermost Cove Beck, but is swept away immediately down the steep slopes.

Despite all these problems it was known the vein was of great value and was rich in mineral, and the mining activities had created much interest, even though much of its success so far had been overshadowed by that of Greenside Mine in the next valley.

The situation was about to change, however, because on the 4th of December 1832 a lease was drawn up for a twenty-one year period, at a royalty of one tenth, and was granted by Mr E W Hasell of Dalemain to the Eagle Crag Mining Co. The man who formed this company was John Errington, a well known miner, originally from Alston and now living in Patterdale.

Shareholders, Eagle Crag Mining Co 1832

John Errington	Patterdale	
George Errington	Patterdale	
William Errington	Patterdale	
John Walker	Patterdale	Blacksmith
William Longrigg	Penrith	Grocer
James Scott	Penrith	Painter
Jas Smith	Penrith	Grocer
John Smith	Penrith	
Mr Irving	Penrith	Innkeeper
George Foster	Penrith	Innkeeper
Robert Clark	Penrith	
Mr Mayhill	Penrith	Attorney
Thomas Atkinson	Penrith	Gent
Edward Heylin	Penrith	Draper
Thomas Harrison	Penrith	Skinner
Thomas Hethrington	Penrith	Gent
George Dixon	Penrith	Innkeeper
Thomas Smith	Penrith	Accountant
Mr Richardson	Penrith	Spirit merchant
John Gale	Penrith	Grocer
Joseph Salkeld	Penrith	Gent
John Harvey		Berrier
Mr Tyson	Penrith	
Charles Scott	Penrith	Plumber
Edward Nicholson	Penrith	Schoolmaster
Henry Nicholson	Thorpe	
G B Hagg	Sunderland	
Thomas Bowman	Hartside	
John Wright	Barton Church	
Robert Hethrington	Maryport	

The company set about working the mine with great vigour, first of all laying a well engineered wagon-way up from the smithy in the valley, where all general work was to be done. The road zigzagged up the southern side of Nethermost Beck, with a branch off for the intended lower workings. The road climbed behind and through the small crags to gain access to a large plateau at OD 1,650 feet, a distance of nearly a mile from the smithy.

It was here the men set to work constructing the final hundred yards of inclined roadway. A large stone house was built to ensure some protection and comfort for these lonely men in their isolated eyrie. In the nearby crags in a natural hollow the gunpowder store was built. The vein here was obvious, and it was the intention of the new company to drive the No.5 Level just under the old workings, where the lead had been plentiful. In an effort to complete this plan, the old workings above were pushed east from No.6 crosscut level and a sump put down forty feet. This became flooded time and again as water poured in from the surface through the open stopes. The water had to be removed by jackroll and bucket, a slow, laborious task and, with the amount of water involved, virtually ineffective. Priority was then placed on driving the No.5 Level the 120 yards or so to come under the workings above, stoping any pockets of ore on the way. Eventually this connection was made.

In anticipation of the ore to be produced, a two tier hand-dressing floor was constructed, and adjacent to this a smithy was built so the drill steels could be sharpened. It is obvious that tons of ore were actually won from these high workings and were dressed here. Still, a major problem must have been the shortage of water for washing and dressing the ore as there are no streams, only water running off the rock face and out of the adit. The magnitude of the construction at this site is difficult to envisage, but the sheer enormity of the project, all those years ago, would have engaged dozens of men and horses over a period of many years.

It was determination and the conviction of the company that brought about a major stroke of good fortune, when a bunch of galena weighing thirty tons was won in 1837. Convinced that the mine was going to provide the shareholders with a bonanza, a lower level, the No.3, was commenced, this being stone-arched for twenty-five yards before joining boulder clay and later, bedrock.

Nearby, another large two-storey stone house was constructed in the shelter of the crag, and this also incorporated a smithy. The large boulder at the rear of the house still bears witness to where the blacksmith tried out the newly-honed drills. Below, the company had built a new dressing floor in expectation of the ore to be gained from the No.3 Level and the ground above. Lower down, two settling ponds were built to prevent waste flowing into Grisedale Beck. The water supply for the new dressing floor came down the open work in the cliff face between the No.5 and No.3 levels. This was also used to transport ore from No.5, and a receiving ore bin was constructed at the base of the cliff just above No.3. From here, it was transferred down more chutes, together with the ore from No.3, for dressing. Along the face from here is No.2 Level at OD 960 feet, cut directly on the vein where the old men had taken a substantial amount of lead and worked out to surface. Shortly, it was planned to work this level. To prevent the level entrance from being smothered by the spoil and scree from the No.3 above, the open stope was choked with huge boulders, thus stopping tons of rubbish filling the level.

After the company had removed all the ore from the upper workings, all attention was concentrated on pushing the No.3 Level forward, again in the hope of gaining the riches of the lead vein which they had experienced in the upper workings. It was also decided to drive the old No.2 Level forward. The No.3 was driven on a distance of at least 150 yards and the men were striking lead and a metallic brown mineral — zinc. At that time zinc had no commercial value and tons of it were brought out and dumped.

The mine was still short of water for dressing, and so to help resolve this a large water leat was cut from Nethermost Beck and directed to just above the newly erected two-tier dressing floor. The ore was sent down the cliff face into a large ore bin, where it was broken with hammers. From here it was slushed onto red sandstone launders, through gratings, and then onto iron plates, where the ore was then crushed down by hand ready for the smelter. Part of the dressing floor was either paved or cobbled, and probably just had wooden walls to afford some protection against the weather. Here also are the remains of another stone building, which was most likely the office. From here the waste from the dressing process was washed down to the settling ponds below.

The No.2 Level followed a thin lead stringer for 160 yards. The whole of the level was railed out with wooden rails topped with metal strips, and was cut in at a fairly steep angle to allow the tubs an easy run out. This must have been some push on return, the tubs being stoutly constructed of sheet iron bolted onto a heavy frame on a wooden chassis with four flanged wheels. At the rear of the tub were two small handles to lift with when emptying the contents, the tipping action effected by means of a pivot on the front pair of wheels. The carrying capacity of these tubs was half a ton.

For the miners driving and mucking out, the air in this level must have been fairly awful because of its small size, and the stench of tallow and the sweat of the toiling men would have added greatly to the discomfort. To compound matters this is a relatively dry mine, and the dust must have created some of the worst conditions imaginable.

In the last few years the company had produced little ore as the drives in both No.2 and No.3 levels had proved virtually worthless. Some lead had been won in No.3 but the rest had been the unwanted zinc.

It must have been in the early 1840s when it was decided that before any more work could be done in No.2 Level a rise should be put up the ninety feet to No.3 to ventilate the mine, and at the same time prove the vein. Sometime in 1845 the rise was completed and it did indeed ventilate the workings, but little or no lead or copper ore was found, and as a result the company had no alternative but to vacate the sett.

The mine fell into dereliction and no further work was done for some years until a group of local adventurers decided to try their luck. These men were James Phillips and Thomas Redfern, both labourers, and Jacob Johnston and John Pattinson, all of whom lived in Patterdale. A lease was granted for the term of three years, with a royalty payment of 1/12th per ton. There was also a clause with regard to poaching and the consequences thereof. The details were finalised on the 7th of January 1862 by Edward William Hasell of Dalemain.

It is impossible to tell what work these men did. It could be that they tried the vein near the beck and drove the No.1 Level, for here the vein is very small and was probably ignored by the early miners who were looking for richer pickings. These were poor men with only enthusiasm and basic mining skills who, perhaps, hoped just to win a few tons of ore, some possibly off the dumps. The level has only been driven thirty-seven yards, and there is no stoping or rise to the few surface trials above. The level, however, has been cut to a good size, being six feet high and four feet wide, which would indicate a working of this period.

The next lease had more credibility, stemming as it did from an enquiry by the Greenside Mining Company and agreed between, on the one hand, John Edward Hasell of Dalemain, and on the other Mr George Head-Head, Mr John Jameson, Mr William Whitwell, Mr Robert Bradshaw Smith, Mr Lancelot Harrison and Mr Charles Stephenson Jackson, on the 6th of August 1872 for a term of one year. The details of the lease were that four workmen were to be employed, the dead rent was to be £15 and that should ore raised be in excess of £150 in value, the royalty would become due at 1/15th for copper and 1/10th for lead.

A similar lease was drawn up with the same covenants on the 6th of August 1876 and signed by the same directors. It seems obvious that the company were just doing exploratory work and did not want the major commitment of a long term lease.

During the period between 1872 and 1876, two major new trials were undertaken. On the western bank of Nethermost Beck, where the Clay Vein crosses, an ancient trial had been tried at surface and some lead was seen. The company decided that this should be tried at depth, and at OD 1,150 feet a smithy was constructed. A few yards of ground cover was taken out and the bedrock soon exposed, and after a few yards of driving a band of slate encountered. The drive continued a while then the project was stopped, and the company turned its attentions to the old No.2 Level.

It may well have been the interest in the zinc or the small lead vein which enticed them to push the level on. From the rise going up to No.3 at 160 yards in from day, the lead vein is two inches wide and the roof taken out to fifteen feet to expose the vein further. This was the last work undertaken by the old men. The level was pushed on, and driven larger in size being six feet high and nearly four feet wide. The miners worked more comfortably here than their predecessors, due to the fresh air created by the water blast from the No.3 Level. The men pursued the level for a total of seventy yards, laying track as they went, nearly up to the barren face but no ore of any value was found.

The management under the guidance of the mine captain, Richard Grenfell, decided that it would be worth driving a level east in order to try and find the east-west vein, which must lie somewhere beneath the towering grey heights of St Sunday Crag. At OD 850 feet the miners first excavated eighteen yards through glacial till, which they stone-arched, in order to secure the level entrance. From this point it was solid rock and the miners were able to continue forward, drilling and blasting until, after fifty-six yards, a vein was intersected going east. It was decided that this should be followed, and the drive was also continued south on the original heading. The driving here was hard going and wheelbarrows were used to clear the waste. Wooden rails were laid from the entrance and on down the eastern heading. After thirty yards the ground became poor as the rock strata altered. Although driving was made easier by the nature of the soft ground, it meant that much of this could not be blasted, but had to be actually hacked out by pick. The place was running with water and it was certainly not the safest of places.

The men pushed on and three cross-stringers were encountered, but these were of no value. The miners had a bait spot at the junction where it was dry, and from where they could see the light from the entrance and smell the sweet fresh air on the wind. Here they could sit for a well-earned break, smoking their clay pipes and probably speculating on the fruitlessness of this trial.

Eventually the company decided enough was enough, having driven a total of 295 yards, 160 yards south in solid ground and a further 135 yards east in very unstable ground, and at no time had they cut any mineral.

Many years later two brothers called Watson, miners from Deepdale, were out walking the fells on the eastern shoulder of St Sunday Crag, when one of them noticed the glint of quartz and gleam of silver — the lost vein. The two miners, realising the value of their find, removed some of the ore and covered the exposed area with grass. The weather was poor, so they made their way back down to the warmth of their homes, discussing the reward they would receive from Captain Borlase, their boss at Greenside. So impressed was he with the ore sample, that he set out immediately with the brothers to examine the vein. Alas, try as they might, neither could locate the spot despite their local knowledge, and from that time on the vein has been known as Watsons' Lost Vein. So ended the mining in the Grisedale valley.

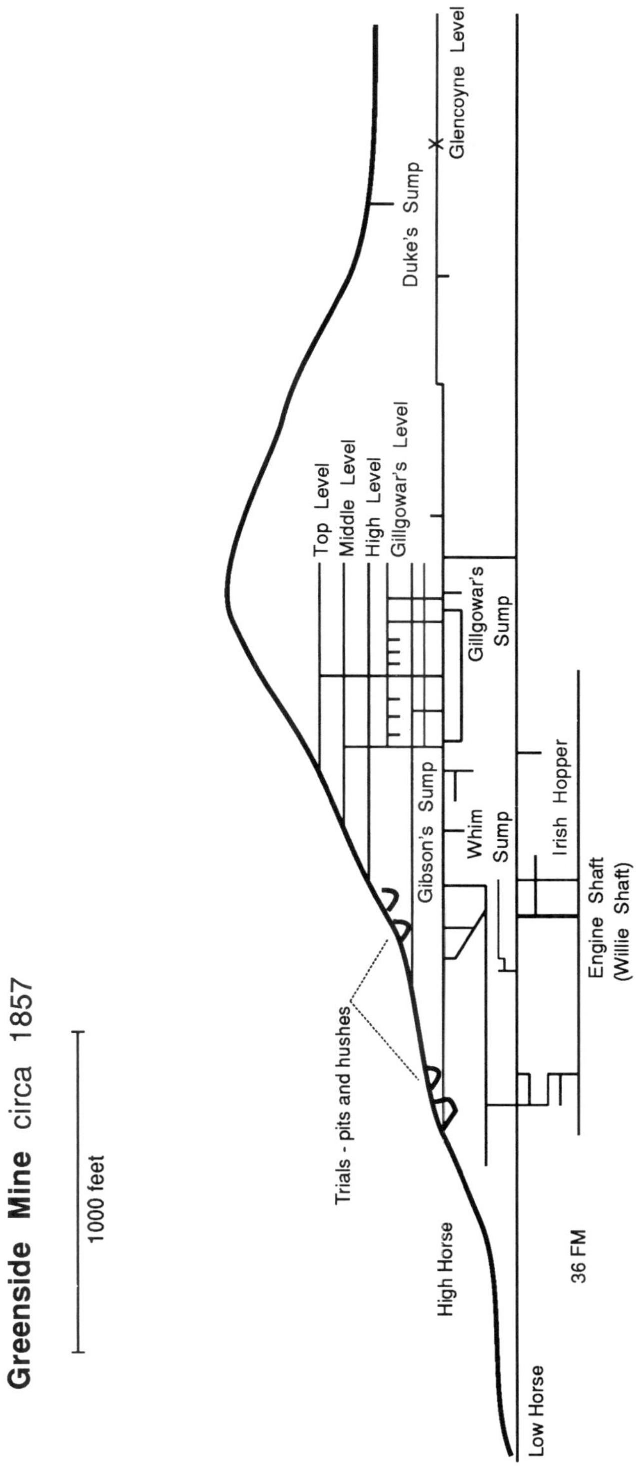
Greenside Mine circa 1857
1000 feet
Top Level
Middle Level
High Level
Gillgowar's Level
Glencoyne Level
Duke's Sump
Gillgowar's Sump
Gibson's Sump
Whim Sump
Irish Hopper
Engine Shaft (Willie Shaft)
Trials - pits and hushes
High Horse
Low Horse
36 FM

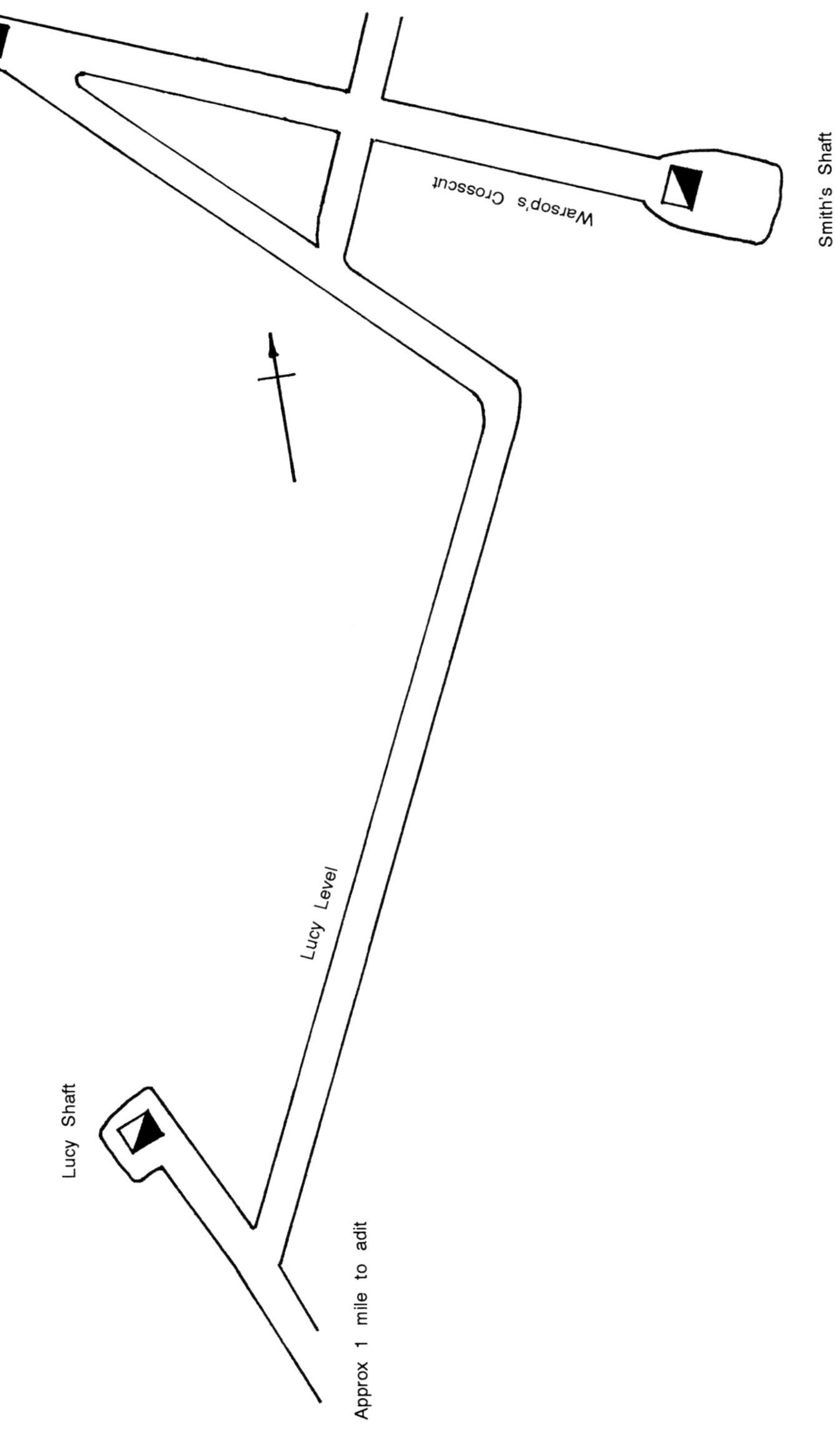

Plan of main shaft workings off Lucy Adit 1885

Greenside Mine circa 1937

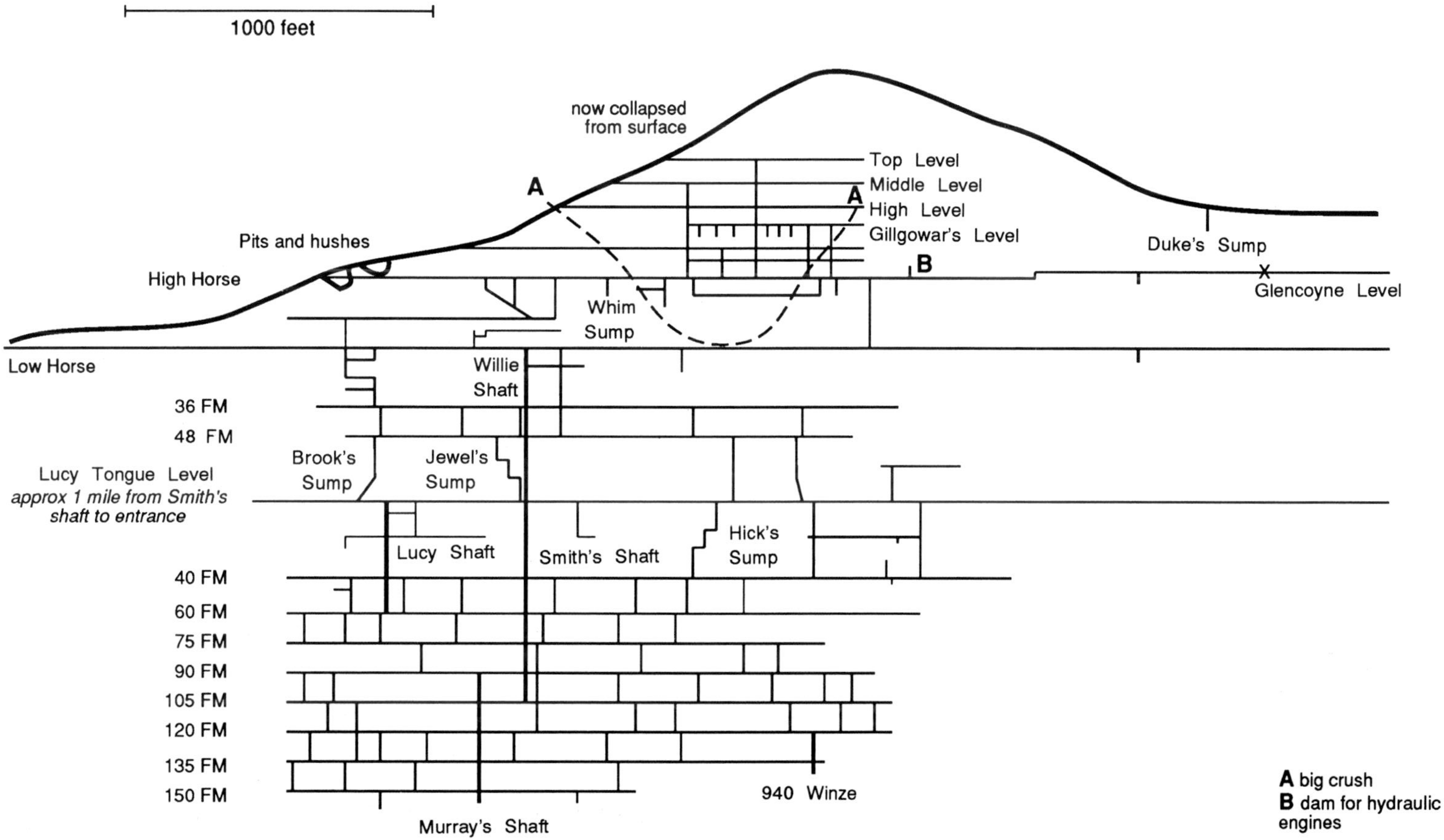

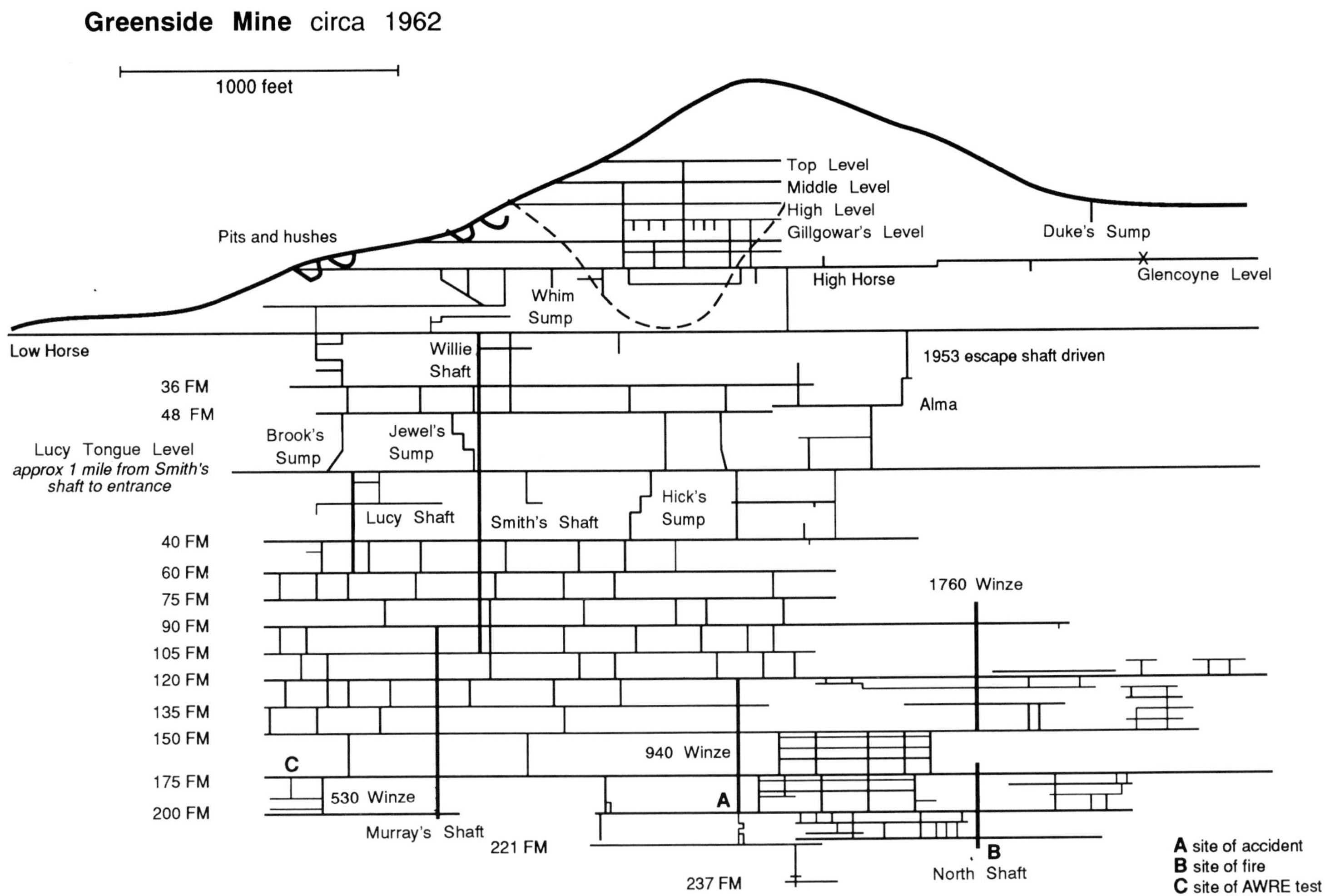
Greenside Mine circa 1962
1000 feet
Pits and hushes
Top Level
Middle Level
High Level
Gillgowar's Level
Duke's Sump
High Horse
Glencoyne Level
Whim Sump
Low Horse
Willie Shaft
1953 escape shaft driven
Alma
36 FM
48 FM
Brook's Sump
Jewel's Sump
Lucy Tongue Level
approx 1 mile from Smith's shaft to entrance
Lucy Shaft
Smith's Shaft
Hick's Sump
40 FM
60 FM
75 FM
90 FM
105 FM
120 FM
135 FM
150 FM
175 FM
200 FM
1760 Winze
940 Winze
C
530 Winze
A
Murray's Shaft
221 FM
237 FM
B
North Shaft
A site of accident
B site of fire
C site of AWRE test

Greenside Mine surface details

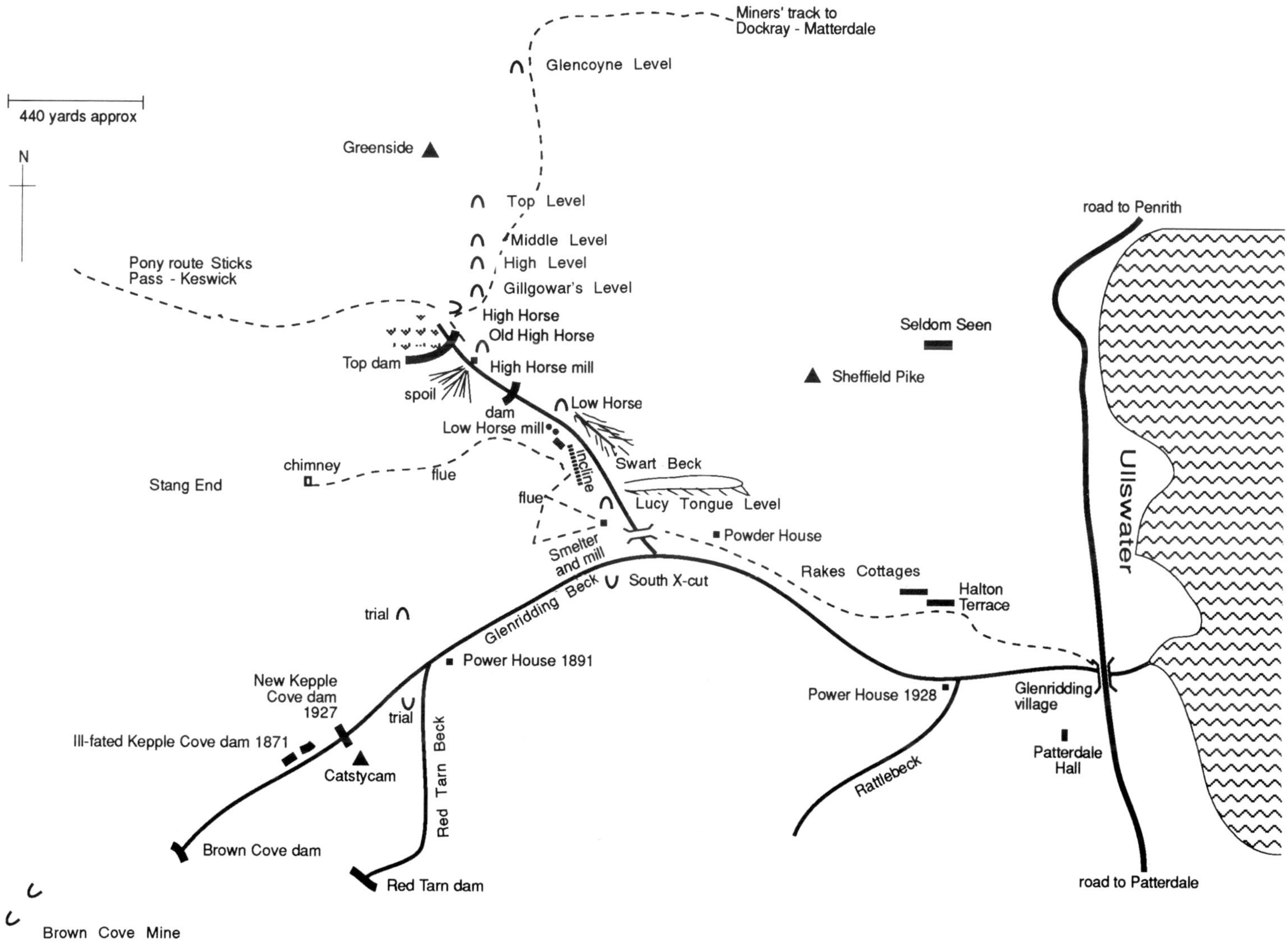

The Geology of Greenside Mine

THE ROCKS of the area are of the Ordovician age and are known as the Borrowdale volcanic series. They consist of bedded volcanic ashes interstratified with lava sheets and tuffs which are chiefly of rhyolite and andesite, in a complex pattern which is intruded by quartz porphyry. This covers an area of many square miles, though in the north and at depth dark Skiddaw slates have been encountered. Much of the rock has a green shade and locally is called "greenstone", which is quite hard and is often associated with rich mineralisation.

The Greenside Lode occupies a fault fissure caused by the faulting of the country rock. Water deposition has, very likely, been the source of the vein filling. The walls are very hard and show no signs of slickensides, the filling being, as it were, frozen to them when first it became deposited. The lode strikes nearly north-south, but in parts varies NNW to NNE, and the hade is eight to eighteen degrees. The strike length is at least 2,400 yards, and is up to 250 feet in width, the vein filling consisting of galena, sphalerite, chalcopyrite, quartz, calcite and barytes.

The vein at surface shows the galena very much disseminated throughout the rock, and appearing as strings varying in thickness from a quarter of an inch to three or four inches. At depth, the ground has been stoped out sometimes up to a width of sixty feet.

Greenside Lode is cut by four veins, the southernmost a strong E-W fault known as the Clay Vein. The main lode has not been identified below this junction. The Clay Vein is sometimes referred to as the Lucy Tongue Vein, simply because this is the vein the Lucy Adit follows.

The second vein, some 200ft to the north, is the Johnson Ross Vein; and thirdly, and most importantly, at a distance of 1,500 feet north, is the North Cross-Course. This hades north whereas the Clay Vein actually hades south. However, the inclination of the Johnson and the shifts of all three are not consistent.

About 200 feet to the north of the North Cross-Course is a branch vein of some significance, but it is only seen above the 175 Fathom Level. The vein filling consists of quartz and mineralised patches of varying richness. The vein is contained within hard silicified walls, alternating with sections of gouge filling where no quartz or sulphides are seen. The walls of the latter sections are soft and rotten, the mineralised sections containing galena and a small percentage of sphalerite, copper minerals, pyrite, calcite and barytes. On surface some of these cross-veins can be seen and examined, as can the main lode.

Topography of Greenside Mine

Name	Ordnance Datum	Circa
Rattle Beck Power Station	650ft	1928
Smelt Mill	1020ft	1835
Basinghall Mill	1100ft	1836
Old Mill	1150ft	1885
South Crosscut	1000ft	1870
Lucy Level	1081ft	1853
Top Power House	1300ft	1890
Low Horse Mill	1580ft	1860
Low Horse Level	1600ft	1835
New Kepple Cove Dam	1650ft	1928
Old Kepple Cove Dam	1750ft	1870
Old Dam Swart Beck	1750ft	1830
High Horse Mill	1780ft	1830
Glencoyne Adit	1790ft	1870
Original High Horse Level	1800ft	1820
Top Dam	1800ft	1870
High Horse Level	1810ft	1827
Duke's Sump	1890ft	1799
Gillgowar's Level	1910ft	1825
Hushes and Trial Pits	1960ft	1750
Brown Cove Dam	2000ft	1870
High Level	2070ft	1800
Middle Level	2170ft	1800
Top Level	2220ft	1800
Chimney Flue	2250ft	1857
Red Tarn Dam	2300ft	1890

Miners Working at Hartsop Hall Mine

1696	J Lowther	N Johnson	R Jopling
1802	T Dodd		
1803	R Tipple		
1830	J Slack		
1832	Mr Bowman		
1834	J Richardson		
1837	L Proud		
1840	J Shaw		
1844	G Nelson		
1845	W Pattinson	J Hodgson	
1850	G Coward		
1855	W Dawson	R Pattinson	G Shaw
1856	T Leck	J Brunskill	
1857	T Milburn	J Milburn	
1861	W Atkinson		
1862	T Watson		
1863	W Hellen		
1864	W Ossables	G Crossthwaite	
1866	W Kilner		
1867	R Thompson		
1868	W Kirk	Mr Whiteside	
1869	Mr Kennedy		
1870	J Brunskill	W Brunskill	P Bainbridge
	R Grenfell		
1875	S Ware		
1878	W Allcock	J Priestman	G Morton
	J Harrison	G Harrison	
1890	G Crossthwaite		
1895	W Harrison		
1905	T Hadwin	Mr Heslep	
1931	T Shaw	J Myles	W J Shaw
	W Hebson		
1940	J H Clay	E Flitters	W Smith
	J Shaw		

The Miners of Greenside

1713	Margaret Vaizee		
1754	G Smedley		
1828	R Hudson		
1829	G Stalker	J Martin	W Welch
1831	J Welch	J Gillgowar	
1832	T Woodhall		
1834	E V Little	W Spark	T Watson Snr
	C Horsley	W Horsley	T Ramsay
	J Gillgowar	J Gillgowar	J Johnston
	J Swinbank	T Gibson	T Hutchinson
	C Eddy	G Walton	J Jacques
	W Millican	James Littel	J Swindle
	C Swindle	W Stephenson	J Ridley
	J Ridley	W Richardson	T Robson
	I Caslow	J Greaves	W Eddy
	J Moffat	T Woodman	J Greenwell
	N Gibson	G Johnston	J Lee
	H Lothian	W Littel	J Featherstone
	W Place	W Patty	W Leck
	R Parker	T Walton	J Holliday
	J Dover	W Todd	T Dawson
	W Keenlyside	J Hymer Jnr	J Millican
	Mr Parker	J Bowman	J Holliday
	J Howard	J Laidlow	W Pattinson
	H Yeats	H Vipond	J Welsh
	L Harriman	J Reed	J Richardson
	J Bainbridge	J Bainbridge	I Shield
	N Vickers	T Alexander	R Brammer
	F Shield	L Potts	M Stanger
	J Scott	J Robson	J Robson Jnr
	N Jack	J Atkinson	C Dixon
	J Millfield	E V Little	T Lec
	H Chapman	F Grahams	G J Little
	H Chapman	W Hymer	J Patty Jnr
	J Patty	J Craig	J Little
	J Hillary	T Teasdale	W Frodsley
	I Hall	J Hodgson	L Fawcett
	H Yeats	J Pearson	J Pearson
	G Little	J Norman	J Nixon
	J Welsh	P Bainbridge	M Ridley

	J Martindale	J Coulthard	J Shield
	J V Little	J Calvert	T W Vipond
	C Shield	W Bird	J Pearson
	L Bell		
1835	J Bainbridge	Mr Greenwell	T Hutchinson
	C Hutchinson Jnr	J Hutchinson Jnr	J Hutchinson Jnr
	N Walsh	Mr Henthorn	T Gibson
	N Jack	W Joseph	C Eddy
	J Redmond	J Mains	T Tindall
	Mr Oglethorpe	W Jackson	R Jackson
	J Jackson	E Little	
1836	G Errington	T Watson	
1837	P Bainbridge	R Alexander	
1838	J Brumbeck		
1839	J Nicholson	N Collins	W Errington
1840	J Pattinson	J Bell	
1841	G Archer		
1842	W Place	Mr Hymer	W Hymer
	J Grieves	G Archer	J Ridley
	T Swindle	Mr Little	G Ridley
	J Patty	Mr Ridley	P Swindle
	Mr Holliday	J Spark	J Pearson
	Mr Hodgson	T Gibson	J Craig
	W D Patty	Mr Shields	W Watson
	T Holliday	T Bird	Mr Castlehow
	G Bewley	J Nanson	J Moffat
	H Vipond	W Nicholson	R Waugh
	G Stalker	W Moffat	F Waugh
	Mr Reed	Mr Parker	A Waugh
	W Bowerbank	Mr Bowman	J Bowman
	G Winder	M Bowman	C Swindle
	R Patty	A Craig	W Irving
	W Leck	M Collin	Mr Brommel
	M Ridley	W Richardson	J Richardson
	Mr Watson	L Henryman	W Vipond
	W Bell	I Harker	G Harker
	Mr Richardson	Mr Greaves	R Browney
	M Lee	J Shield	J Gillgowar
	E Lee	Mr Keenlyside	E Robson
	J Patty Jnr	J Grieves	J Harker
	R Patty Jnr	J Rucastle	Mr Slack
	Mr Row	R Parker	W Todd
	J Johnson	A Moffat	W Todd Jnr
	W Patty Jnr	W Place	T Swindle
	W Welch	Mr Muse	T Milburn
	J Coulthard	Mr Coulthard Jnr	H Rucastle
	P Bainbridge	J Hutchinson	C Eddy

	J Gillgowar	R Malone	J Vickers
	Mr Watson	W Elliot	W Tyson
	J Thompson	W Mews	T Hymer
	T Robson	R Thompson	G Little
	G Hindmoor	J Coward Sm	
1843	S Bird	T Lowes	Mr Houston
	F Craig	Mr Walton	T Watson
	G Nelson	W Richardson	J Richardson
	L Merryman	W Vipond	G Archer
	J Ridley	G Ridley	I Little
1844	W Errington	H Errington	J Nanson Jnr
	J Whitfield	W Crosbie	R Woodburn
	W Nicholson	J Chapman	
1845	L Harrison	T Milburn	R Pattinson
	W Dickinson	R Thexton	J Gill
	T Wallace	R Hilson	T Hilson
	J Fothergill	P O'Brian	J Bell
	W Harrison	J Birkett	T Mackintosh
	L Bell	V Swindle	I Castlehow
	T Watson	J Little	E Poole
1846	R Woodburn	J Vickers	W Brunskill
	W Langstaff	J Parker	
1847	J Bateson	J Hill	J Wilkinson
	W Bateson	J Douglas	
1848	J Tallentire	T Lowes	L Dowies
	J Airey	W Tyson	
1849	J Scrivens	C Watson	
1850	J Mains	W Lancaster	
1851	Mr Oglethorpe	J Martin	P Mason
	J Bennett	T Bennett	
1852	J Nixon	H Weightman	L Fawcett
	A Poll	W Carruthers	T Hilton
	W Watson		
1853	R Wilson	J Gilchrist	J Stanger
	J Budd	Mr Cookson	R Alexander
	W Bell Jnr	J Bell	T Milburn
1854	J Duglas	G McCall	J Thwaite
	J Shields	J Barker	J Brunskill
	F White	G Swindle	
1855	G Ramsay	C Hethrington	A McMillan
	L Hodgson	C Taylor	C Abbott
	J Glenwright Sm	W Muse	J Thompson
	J Shields	P Swindle	T Sibson
	Mr Johnson	J Murray	
1856	G Ramsay	G Newton	W Seatree
	D Pattinson	T Plaskett	
1857	J Vickers	W Smith	J Place

1858	R Millican	J Robinson	A Reid
	J Place	F White	J A Craig
	J W Parker	J R Fothergill	E Cherry
	J W Shields	T Gibson	M P Millen
1859	W Walton	G Harker	D Bellas
	E Lee		
1861	F Gould	F Gould Jnr	W Harris Jnr
	J Harris	J Hancock Jnr	R Hancock
1863	T Nanson	J Stobbart Jnr	J Stobbart
	J Johnston		
1864	C Meadway Jnr	C Meadway	T Lightburn
	J Jackson Jnr	W Jackson	G Archer
	C Mealbury	J Millican Jnr	T Millican
1865	W Armstrong	T Brooks Jnr	T Brooks
	D Douglas	W Armstrong Jnr	B Grisedale
1866	P Robinson	R Hayton	
1868	J Dewis	W Harrison	T Robinson
1869	D Hodgson Jnr	J Hodgson	W Jackson
	G Brumwell	J Compston	G Nelson
	G Hethrington	R Nicholson	T Coward
	J Coward Jnr		
1870	W E Hastings	R Paul	J Metcalf Jnr
	J Metcalf	W Mathison	W Stobbart
	J Norman Jnr	N Norman	A Millican Jnr
	J Millican	J Pattinson	B Hethrington
	J Atkinson	J Rucastle	T Moor
	D Allison	J Weightman	T Nixon
	W Pattinson	J Mattinson	T Dai
	W Nicholson	T Walker	T Coward
	T Raw	J Compston	J Coward
	T Salkeld Shp	D Martin Hk	G Hacker
	J Foulston	J Robinson	J Brumwell
	J Scott	J Hewitson	J Harrison Sm
	J Brownrigg Sm	B Grisedale Sm	J Harrison Sm
	J Nicholson Sm	G Wear Sm	B Dixon Jnr Sm
	J Robinson Sm	W Nanson Sm	J Lancaster
	W Potts Jnr	T Coward Jnr	D Hodgson
	T Sowerby	T Sedgewick	W Sedgewick
	R Bainbridge Sm	J Curry	W Jackson Sm
	D Wilson Sm	J Graves Sm	J Nelson Sm
	T Whinn Sm	T Fenton Sm	T Leck Sm
	E Taylor Sm	G Harrison Sm	J Thompson Sm
	G Coward Sm	B Dixon Sm	J Little Bs
	J Nicholson Bs	J Wensley Bs	W Walker
	T Moor		
	I Peart	R Thexton Hk	J Wear Sm
	J Reynolds Sm	W Nelson Sm	J Nixon Sm

	W Dixon Sm	J Thexton	G Bell
	I Nicholson	W Howgill	T Boustead Ms
	J Boustead Ms	D Allison Ms	D Brown Jr
	M Allcock Jr	G Mitton Jr	J Bainbridge Jr
	J Brown Jr	T Watson Bs	J Leck Bs
	G Watson Bs	J Bownass	G Ridley
	G Newton	H Bateson	J Pattinson
	T Robson	E Pool	W Nelson
	J Pool	C Robinson	W Nanson
	T Metcalfe	J Castlehow	W Glenwright
	J Watson	J Douglas	T Brooks
	T Redfern	W Jackson	G Bell
	J Atkinson	J Millican	G Ramsay
	J Lake	J Place	J Spark
	A Watson	W Smith	J Norman
1871	T Brooks	J Mattinson	J Douglas
	J Lake	G Brumwell	T Hutchinson
	B Bolt	M Shaw	J Bownass
	W Smith	J Crompton	G Ridley
	G Newton	W Place	J Wilson
	G Hodgson	T Metcalfe	G Castlehow
	W Elland	J Richardson	J Taylor
	W Walker	J Watson Bs	J Nicholson
	J Little	T Robinson	W Brunskill
	W Harrison	J Stewart	H Barton
	D Wilson	J Bowman	F Graham
	R Thexton Hk	T Darghue	E Darghue
	J Greaves Sm	W Jackson Sm	T Winn Sm
	W Sedgewick Sm	J Thirwell Sm	J Dixon Sm
	W Dixon Sm	C Hethrington	J Wear Sm
	W Martin Sm	J Robinson Sm	J Robinson Sm
	J Brownrigg Sm	W Potts Sm	J Pattinson
	T Dodsworth	W Watson	V Swindle
	R Grenfell	J Oglethorpe	T Taylor
1873	W Johnston	M Shaw	
1874	J Armstrong Jnr	J Cole Jnr	J Brown Jnr
	C Jewell Jnr	J Kilner Jnr	W Kilner Jnr
	W Lake Jnr	R Lake Jnr	J Murray Jnr
	E Pool Jnr	W Pattinson Jnr	R Shaw Jnr
	J Wilson Jnr	W Robinson Jnr	J Milburn Jnr
	J Robinson Jnr	J Robinson Jnr	W Harrison Jnr
	D Harrison Jnr	J Peet Jnr	W Peet Jnr
	J Harris	H Harris Jnr	T Rigg
	W Kilner	J Nicholson Jnr	J Nicholson
1875	T Graham Jnr	J Graham Jnr	W Murray Jnr
	J Peet Jnr	W Robinson Jnr	T Graham Jnr
	H Grenfell Jnr	H Grenfell Jnr	S Gilbert Jnr

	W Gilbert Jnr	J Harris Jnr	W Layland
	E Layland Jnr	J Currie	J Scott
1877	R McKendrick	J Millican	G Shaw
	T Sewell	R Hooper	T Hindmoor
	J Jackson	T Hutchinson	G Cookson
	J Clark	H Morton Jr	J Oglethorpe Jr
	J Ryan Bs	J Oglethorpe Bs	J Brown Bs
	J Little	J Nicholson	J Wilson
	J Miller	G Ridley	G Newton
	W Place	J Dickinson	J Dickinson
	M Pattinson Ms	R Shaw Sm	R Shaw Sm
	W Murray Sm	W Brunskill Sm	J Brunskill Sm
	J Reynolds Sm	S Murray Sm	J Coward Sm
	J Cowper Sm	J Curry Sm	G Harrison Sm
	J Harrison Sm	J Lancaster Sm	J Bowman Sm
	W Harris	H Harris	W Maughan
	J Cowan	T Robinson	T Cookson
	J Blamire Hl	J Weightman Hl	J Martin Hl
	W Shaw	R Grenfell	P Millican
	J Lake Jnr	J Lake	J Oglethorpe
	D Newbiggin	J Brunswick	W Mattinson
1879	J M Harris Jnr	T Harrison Jnr	S Jewell
	T Jewell Jnr	T Kilner Jnr	W Kilner Jnr
	W Lancaster Jnr	T Millican Jnr	W H Mitton Jnr
	J Murray Jnr	W Nelson Jnr	G Nelson Jnr
	J Oglethorpe Jnr	W Pattinson Jnr	W Pattinson Jnr
	G Slater Jnr	W Watson Jnr	J Walker Jnr
	H Wall Jnr	E Lancaster Jnr	J Williams Jnr
	J Williams	G Mitchell	T Graham
1881	T Bray Jnr	J Bray	J Watson
	J Robinson Jnr	G Robinson	W Hicks Jnr
	T Hicks	D Pattinson	G Hethrington
	R Hethrington		
1883	J Jackson	R Brown	T Bellas
	R Stout	D Pattinson	J Atkinson
	G Sharp	W Chugg	D Bellas
	C Robinson	T Sewell	J Denis
	J Bownass	B Bolt	J Smith
	J Peart	J Lake	J Watson
	W Walker	T Metcalfe	J Thexton
	E Lancaster	J Pool	J Wilson
	G Swindle	E Pool	G Newton
	W Place	J Dickinson	J Dickinson
	J Hethrington	W White	T Kilner
	J Edmondson Bs	J H Millican Bs	W Dalton
	J Kirkup Hk	R Hayton	J Millican
	R Shaw	R Shaw	J Edmondson

	E Thompson	R Mounsey	W Nelson Sm
	H Wall Sm	B Grisdale Sm	J Harrison Sm
	W Harrison Sm	W Sewell Sm	W Harris Sm
	M Brumwell Sm	J Murray Sm	J Wilson Cl
	J Smith	W Harris Frm	J Robinson Frm
	G Brown Hl	G Swindle Hl	J Cooper Hl
	J Blamire Hl	T Winn Hl	J H Horn Jnr
	J Pool Jnr	W J Swindle Jnr	J Bainbridge Jnr
	T Jackson	G Castlehow	G Hethrington
	G Coulthard	G Shaw	J Shaw Jnr
	J Shaw Jnr		
1884	T Pattinson Jnr	J Nelson Jnr	J Crossthwaite
	J Pattinson Jnr	T Shaw Jnr	T Hogarth Jnr
	J Murray Jnr	G Brown Jnr	J G Hayton Jnr
	G Pattinson Jnr	T Castlehow Jnr	T Bainbridge Jnr
	J Wall Jnr	A Murray Jnr	D Brown Jnr
	G Robinson Jnr	W Jackson Jnr	H Rucastle Jnr
	T Curry Jnr	L Pattinson Jnr	W Jewell Jnr
	W Chappelhow	W Neadham Jnr	J Shaw
	W Sedgewick	W Sedgewick Jnr	T Milburn Jnr
	W Milburn	J Stockdale	
1885	W Chugg Jnr	R Chugg	W Nelson
	T Bellas Jnr	T Bellas	
1886	S Cutliffe	R Shaw	G Shaw
1887	J Peart	W McGhie	J Corner
	R Brown	J Birkett	W Hicks
	J Lake	R Stout	G Shaw
	T Metcalfe	W Hugill	T Jackson
	J Boot	J Wilson	W Place
	J Nicholson	T Bray	M Pattinson Ms
	W Robinson Ms	T Bainbridge Jr	J Edmondson Bs
	G Brown Bs	W Dalton	J Ryan
	J Jackson	R Lake	W Barron
	G Barron	W Murray Jnr	J Murray
	W R Mitchell		
1889	J Wall Jnr	J Darque Jnr	J Davis Jnr
	R G Hayton Jnr	R Hayton Jnr	J Darque Jnr
	J Lancaster Jnr	R W Wilson Jnr	T Watson Jnr
	T Coward Jnr	J Hogarth Jnr	T Davis Jnr
	G Lancaster Jnr	G W Brown Jnr	T Rigg Jnr
	J J Hind Jnr	E Bainbridge Jnr	J Cookson Jnr
	D Coward Jnr	G Robinson Jnr	T L Marr Jnr
	J Marr	G McGhie Jnr	J McGhie
	W Brown		
1890	R C Hayton Jnr	J Murray	
1891	T Pollock	R Slee	J Wall Jnr
	H Wall	T Metcalfe Jnr	J Metcalfe

	J Corner	W Hogarth Jnr	R Hogarth
1894	T Pattinson Jnr	L Pattinson	I Roberts
1896	Mr Murray	Mr Rigg	I Robinson
	J Barker Jnr	R Barker	G Newton
	R Ham	J Sedgewick	W Harris
	G Shaw	T Shaw Jnr	
1897	T A Terrill Jnr	T A Terrill	J Anderson
	H Rucastle Jnr	H Rucastle	
1899	E Parker Sm	J Parker Jnr Sm	J Coward Jnr
	T Coward Jnr	T W Dewis Jnr	J Dewis
	J Coward Sm		
1900	W Lanning	W J Lanning	T Pod
	W Steally	T Lake	R Lake
	J Wilson	F W Hind	T Shaw
	R Shaw	J Shaw	W Rudd
	J Rudd	W Rudd Jnr	I Rudd
	W Nanson	J Nicholson	T Watson
	T Pattinson	R Shaw	W Hayton
	P Kearns	T Weightman	J Birkett
	W Shaw	T Greenhow	W Lawrence
	W McGhie	P McGhie	C McGhie
	T McGhie	T Milburn	E Borlase
	W Hicks	I Roberts	R Mallett
	W Hocking	T Murray	T W Lancaster
	G Weightman	J Watson	R Stout
	D Hicks	T Lishman	T Dawes
	J Hodgson	D Brown	W Chugg
	W Pool	W Coward	T Coward
	A Murray	G Cookson	W Mossop
	H Sullivan	J G Hayton	W Lawrence
	T Greenhow	G Hethrington	T Bray
	R Bray	E Rowlands	W Griffiths
	D Moyle	T Jones	J Jones
	D Evans	J Murray	D Boward
	G Thompson	C Jeffreys	
1901	W Brumwell	T Graham	H Graham
	W Lancaster	G Lancaster	W Hocking
	Mr Jones	J W Lancaster	G Cookson
	D Coward	J Parker	
1902	J Rundale Jnr	T Rundale	R Harrison Jnr
	J Harrison	J R Dewis	E Bainbridge
1903	T Lampohire	F Mitchell	R Mallett
1904	W R Chugg	T Davis Jnr	T J Davis
	G Castlehow	W Pitchford	P Millican
	J Lishman	W Shaw	
1905	J Lawrence Jnr	A Lawrence	J Cooper Jnr Sm
1907	G Ashwood	J Cooper Sm	J Peart

1908	F J Dewis	J Kyles	
1909	S Guardhouse	J Cole	A Cole Jnr
1910	A Cummings		
1911	S Jones Jnr Sm	J Jones Sm	G Voletti Jnr
	R Brown	J Norris	W Greenhow
	W H Borlase	E Roach	
1913	W E Cross	T Nanson	
1915	A B Coward	C Pala	
1916	E Parker Sm	J Parker Sm	
1917	N Kirkup	R H Bell	
1919	J S Codling	B E Jones	
1920	F J Stoddern Jnr	F J Stoddern	J Furze
1921	D Brown	J Williams	
1922	T Graham	J H Holliday	J W Pattinson
1924	W J Welch Jnr	J Welch	
1925	A Mignonelli	C Pala	L Largioliah
	G Webster	W Brown	B Jones
	G Volletti	S Webster	J Calvert
	J Bouch	J W Birkett	I Coward
	J Brown	J Hind	T Graham
	F Dawes	J Welsh	G Hall
	F Mitchell	T Robinson	R Borrodale
	T Lister	G Robinson	J Kelt
	T H Hodgson	A Mumberson	L Burnett
	T Hind	J Murray	W Murray
	E Cox	C Jenkinson	H Birkett
	J N Hays	F Rivers	G Birkett
	G Jones	T Grenfell	W D Comer
	O Comer	T Bainbridge	J Slieghtholme
	A G Barclay	O Comer	J Oglethorpe
	J Brown	H Oglethorpe	P Dawes
	G Brown	J Wilkinson	J Sedgewick
	G Ashwood	W D Watson	P Castlehow
	W Pitchford	A Pitchford	J Banks
	J Hebson	F Todd	J Nicholson
	I Blamire	W Armstrong	W Mitchell
	H Watson	H Holliday Hk	J W Wynn Hk
	M J Hayton Cl	G Walker Frm	H J Millican
	T Graham	P Troughton	J Banks
	W Pitchford	G Harrison	E Bainbridge
	A Routledge	J Nicholson	F Todd
	A Pitchford	J Hebson	M Place
	H Grenfell	G Webster	W Richards
	M S Place	G B Mackenzie	
1926	J W Birkett Jnr	J Birkett	H Stratton
	W J Turner	J H Holliday	J H Storey
	W Richards	J B Graham	

1928	W Hind	W Pitchford	A Pool
1929	G E Stewart	J Hamilton Jnr	J Hamilton
1930	I Blamire	O Hind	G Cole
	W Richards	G Harrison	F J Harris
	J Pattinson	W Deswood	W J Richards
	J J Park	J Dawes	
1934	M Liddle Jnr	R Liddle	W E Cross
1936	W T Shaw Jnr	T Shaw	J Connor
	C Connor Jnr		
1938	R Ackerley Jnr	J W Ackerley	T Graveson Jnr
	G Graveson	R Slieghtolme	J E Cross Jnr
	J Deeney	J Pool	A Graham
	Mr Harry Mgr	Mr Friend Mgr	Mr Purvis Mgr
	Mr King Mgr		
1940	J Barnsley	W Little Jnr	W Little
	J R Tallentire	A G Tallentire	J Jenkinson
	W Hebson	D Flynn	
1941	T Wilson		
1942	W Turner	J J Park	W Brown
1943	A A Mignonelli	W C Wilson	
1945	T E Hope		
1946	A Lewis	M C Park	B Corlett

Miners Nicknames of Around 1948

J Brown (Jackets)
G Cole (Charge Stick)
H Birkett (Perry)
W Richards (Primer)
P Richards (Basher)
Unknown (Tomato)
C Jenkinson (Old Clarry)
A Pitchford (Mike)
E Turner (Tamma)
R Shaw (Curly Bob)
J Pattinson (Finkle)
J W Wynn (Endboard)
H Wynn (Bizzler)
D Gilcross (Dr Doolittle)
J Hope (Tot)
I Bennett (Ike)
W Brown (Twigs)
I Coward (Thobbut or Ike)
J Birkett (Jopey)
H Richards (Sally)
T W Holyland (Toss Willy)
W Jenkinson (Pym)
Percy Dewes (Dolly)
J Pitchford (Pap)
W Murray (Captain)
J Oglethorpe (Locks)
A Santamara (Santa)
Fred Wynn (Fearless)
A Blamire (Fanny)
J Harrison (Flowery)
J Hind (Scan)

1948	J Watson	J Watson	W Burnett Elc
	J Paterson	J Wilson	J Smith Bs
	T Kemp	T Hodgson	T Hodgson Jnr
	T Bainbridge Jr	Mr Andrews Jr	J Coward Jr
	R Corlett	J Rohde	J Browe

	G Birkett	J W Birkett	H Oglethorpe Fm
	J Kilgower	J Varey	J Bardsley
	T Hind	H Birkett	W Murray
	R Borrowdale	D Cole	J Cole
	J Coulson Ftr	J Hewer Jnr	J Hewer
	J Stewart	T Doughty	J Kyles
	R Jenkinson	W Jenkinson	F Hadwin Ftr
	R Hind	T Hope	J Hope
	L Mulryan	T Mulryan	D Winder
	J Read	Mr Fisher	Mr Robinson
	F Hadwin	W Patrickson	J Kirkup
	J Hewer Jnr	F Howett	E Turner
	W Tallentire	W Tallentire Jnr	A Tallentire
	J Tallentire	J Gilchrist	H Birkett
	O Pitchford	I Coward	J Wilson
	W Wilson	J Hamilton Frm	J Dawes
	F Dawes	D Wilson	J Blair
	J Barnsley	H Taylor	B Jenkinson
	J Teasdale	R Pitchford	L Mullryan
	R Mallison	J Miller	G Gibson
	A Brown	A E Sinkinson	T E Hope
	J Kemp	T Charlton	L Largioliah
	E Pool	J Park	D Hodgson
	W Hicks	J Coulson	A Nelson
	H Taylor	R Borrowdale	W Curry
	E Dawes Elc	T Hodgson	A Blamire
	J Hamilton	A Cooper	T Hodgson
	J Hill	T Mortimer	H Holliday
	T Simpson	H Bruce	J Ladyman
	Jock Young	A Boustead	E Turner
1950	J Kirkup	R Hind	I Coward
	A H Blamire	Mr Place Off	H Evans Off
	Mr Trembath Off	J Brown Off	M Robinson Off
	Mr Champion	Mr Dennis Sur	J W Hicks Str
	G Fleming Str	Mr Robinson	Mr Williams Lab
	R Dixon Ml	R Sleightholme	G Gibson Ml
	F Dorrien	W Watson	J Park
	Mr Miller Elc	A Cooper Elc	A Nelson
1956	W H Holliday	R Denwood	J Teasdale
	S J Sowerby	L Dixon	M A Mignonelli
	J Hamilton	R Deerwood	M Mignonelli
	L Mignonelli	Mr Dickey	O Jenkinson Off
	C Connor Mgr	C Thompson Off	

The above miners' names are taken from various sources, marriage, birth and death certificates, company documents and personal recollections. As a consequence some names may have been duplicated over a period of years, however this number is very small. The duplication of names and also initials usually means that more than one member of a family was employed at the mine at the same time or over a period of years. Large families were the norm and it was traditional for sons to follow their fathers into a profession.

It has not been possible to list all workers or their trades.

Abbreviations

Junior	Jnr
Manager	Mgr
Foreman	Frm
Blacksmith	Bs
Hauler	Hl
Surveyor	Sur
Storeman	Str
Joiner	Jr
Fitter	Ftr
Horsekeeper	Hk
Cleaner	Cl
Millworker	Ml
Electrician	Elc
Smelter	Sm
Mason	Ms
Labourer	Lab
Office	Off
Shop	Shp

Greenside Production Figures

year	tons of lead	tons milled	£ value	silver ozs
1838	764	995	15280	8978
1839	453	588	8607	8077
1840	792	1029	14256	7726
1841	642	834	12840	7413
1842	662	860	11916	7510
1843	739	960	12563	10655
1844	759	986	12903	14709
1845	1339	1740	24102	17525
1846	1157	1504	21983	17554
1847	902	1172	17138	13153
1848	819	1064	13923	15288
1849	1718	2405	23979	21369
1850	1770	2478	20176	24473
1851	1633	2286	29320	29665
1852	1233	1800	20933	22194
1853	1212	1696	28675	16850
1854	1164	1629	23299	19021
1855	1325	1855	29257	23890
1856	1114	2004	25346	18968
1857	1273	1654	23450	19242
1858	1527	1985	32523	19250
1859	1268	1711	27237	19335
1860	1229	1659	27889	18515
1861	1327	1990	15548	19859
1862	2202	2906	40319	20173
1863	1651	2261	35616	19293
1864	1447	1938	32142	19702
1865	1580	2117	32577	24844
1866	1633	2185	36747	23874
1867	1533	2360	32607	22264
1868	1474	2211	30143	23601
1869	1620	2430	32031	25035
1870	1717	2644	33556	25970
1871	1557	2086	28457	22695
1872	1293	1732	26104	19696
1873	1107	1605	24001	18897
1874	1097	1590	22962	15512

1875	1211	1816	26897	18731
1876	1334	1787	27968	19649
1877	1327	1897	25898	19706
1878	1289	1727	22727	18054
1879	1236	1767	18744	15515
1880	1241	1712	19699	16178
1881	1095	1511	14930	13746
1882	1157	1543	16254	15161
1883	1020	1366	13311	11096
1884	1108	1529	12266	15545
1885	1329	1834	13801	13495
1886	1553	2134	17514	11924
1887	1438	2002	18654	13860
1888	1286	1774	18431	16487
1889	1352	1865	19237	15232
1890	1102	1608	14172	10233
1891	1336	1950	14467	13498
1892	1485	2155	17029	14419
1893	1255	1757	12293	13273
1894	1300	1820	12319	12725
1895	1339	1714	15696	13271
1896	1117	1474	12480	14732
1897	1258	1610	13996	14432
1898	1396	1842	18811	14732
1899	968	1277	13410	10702
1900	1123	1483	15521	11656
1901	1224	1615	16472	10839
1902	1163	1535	13508	10385
1903	1284	1694	13762	2120
1904	1203	1588	14456	1330
1905	1253	1729	17686	*16470
1906	1178	1507	20394	*14450
1907	1210	1548	21691	*11120
1908	1157	1480	14695	*13900
1909	1269	1624	15176	*15360
1910	1310	1676	17007	*18660
1911	1697	2172	22050	*18210
1912	1385	1772	21814	*15020
1913	986	*1202	17011	*10800
1914	750	*960	13454	
1915	521	*666	12031	
1915	18	bought from Threlkeld Mine		
1916	634	*811	15978	
1917	662	*847	16081	
1917	167	bought from Threlkeld Mine		
1918	750	*960	16672	
1919	586	*750	13238	

1920	909	*1163	26003	
1921	397	*568	9925	
1922	670	*857	12181	
1923	629	*792	11694	511
1924	752	8805	18197	719
1925	866	10840	21562	577
1926	1028	12740	21484	541
1927	1500	14640	22639	359
1928	254	2660	3607	90
1929	*1460	12150	11236	
1930	*1465	12103	11000	
1931	*1841	12217	18893	
1932	*1659	12707	13692	
1933	*1278	12323	8761	
1934	*1061	9470	7999	
1935	*752	8805	4960	
1936				
Total	116130	217043	7384871	1222288
1937	1650	*15000	37950	
1938	2083	23150	33328	
1939	2217	22420	*110850	
1940	7680	78675	*384000	
1941	*6834	*70000	*341700	
1942	6638	71540	*331900	
1943	3212	*35699	*160600	
1944	3170	*33000	*158500	
1945	2450	*25000	*122500	
1946	2405	*24500	*120250	
1947	2489	24332	*124450	
1948	2238	22649	*111900	
1949	2166	22966	223765	
1950	2121	23573	223765	
1951	2159	23989	349758	
1952	*1966	21849	269342	
1953	*2071	23018	188461	
1954	*2337	25967	224352	
1955	*2070	*23000	219420	
1956	*2070	*23000	149040	
1957	*1800	*20000	129600	
1958 AWRE				
1959 AWRE				
1960 AWRE				
1961	*1620	*18000	121500	
Total	268,888	1,105,413	£11,521,802	1,222,288

* Denotes estimated figures where a bulk figure is known for the period. In a few instances this is an educated guess in line with production during that period. The total amount of ore milled and lead and silver recovered will never be known accurately, but many of the above figures are from the original account books and can therefore be relied on.

It is interesting to note that the recovery of the lead is much higher in the early period than later. This is due to the fact that the selection of ore being fed into the crushers was virtually hand picked. Further to this, lead was obviously being produced well before 1838 for when the main batch of figures are available. From 1822 an average of 800 tons per year of ore treated would increase the overall total by 12,800 tons. It is important to remember, however, that these figures are for treated ore and not for the vast amounts of dead waste from barren driving, much of which is still to be seen lying in the enormous waste heaps at High Horse, Low Horse and Lucy Tongue levels.

Glossary of Mining Terms

ADIT — A horizontal tunnel driven into the hillside for access to the vein or to drain water from the mine.

BACK FILLING — Waste rock which is dumped in a disused area of the mine to save the time and expense of bringing it out to day.

BACKS — Workings or vein directly above the level.

BAIT — Miners lunch or break.

BALL MILL — Revolving steel drum containing steel balls for crushing the ore.

COFFIN LEVEL — A narrow tunnel, coffin-shaped in cross-section, usually dating to the Elizabethan period and driven without the use of explosives.

COLLAR — The area around the top of a shaft.

COUNTRY ROCK — The rock in which the vein is found.

CROSSCUT — A tunnel driven through country rock to cut the vein and often at right angles to the vein.

DAY — Underground term meaning "surface" or "outside".

DEADS — Waste rock stacked neatly to support the mine walls.

DRIFT — Tunnel or level along the vein.

FATHOM — Six feet.

FLOTATION — A system to separate the concentrated mineral by the use of oil rising through water, to which the concentrate has adhered.

FOOTWALL — The underlying wall of a stope or sloping vein.

GANGUE — Worthless minerals, usually quartz and calcite, making up the mineral vein.

GALENA — A sulphide of lead.

GRIZZLEY — A series of iron bars set apart to size the ore.

HADE — The inclination of a vein or strata.

HANGING WALL — The overlying wall of a stope or sloping vein which tends to form the roof of a working.

HOPPER — A wooden chute fixed into an aperture in the base of a stope so that ore collected above can be discharged into ore tubs in the level below.

HORSE LEVEL — A main adit level where horses are used for underground haulage.

HUSHING — A method by which dammed water was released down the fellside, the force of which removed the surface grass and shale and thus exposed the vein.

INCLINE — A tunnel driven at an angle, usually between 33 and 50 degrees from the horizontal, or a tramway for lowering ore down a slope either on the surface or underground.

JIGS — Machines for separating mineral and waste by means of agitation and water.

JUMPER — Hand-held drilling steel, various lengths.

KIBBLE — A bucket-like container for hauling ore and equipment up shafts.

LEATS — Courses cut into the fell to channel water for the operation of waterwheels etc.

LEVEL — Horizontal tunnel giving access to workings, also providing drainage.

LODE — Vein carrying mineral.

OLD MEN — Early miners.

ORE PASS — A shaft or chute made of timber or stone, or cut into solid rock and used for passing ore and waste rock to the level below.

PARCEL — An amount of mineral dressed ready for selling.

PELTON WHEEL — An enclosed wheel with cups which accept a powerful jet of water. From the drive shaft of the wheel machinery of all sorts can be driven.

PILLAR — A part of the vein which is left in position to support the walls of the mine.

PORTAL — The entrance to a level, usually constructed from local materials.

RISE — An internal shaft put upwards in the vein, which may eventually connect with the level above. If laddered it is then referred to as a manway or winze.

ROYALTY — An area of land determined by a lease, or the payment demanded by the mineral owner on every ton of milled ore produced by the operator.

RUN-IN — When the roof or sides of a level collapse.

SETT — The area of land determined by the lease and on which the operator has freedom to mine.

SHAFT — A hole nearly always vertical but can be inclined, whereby ore is winched from one level to another and by which the miners gain access to the workings.

SLIMES — The fine slurry from the mill created in the crushing process.

SOLE — The floor of a level.

SPOIL — Waste rock discarded outside the mine entrance, usually with no or little mineral content.

STAMPS — A series of large iron hammers for breaking up the ore, usually driven by a waterwheel.

STOPE — An area where the ore has been removed - the working part of the mine.

STRINGER — A very narrow vein.

SUMP — A shaft sunk in the sole of the level.

TAKE-NOTE — A licence to search for minerals.

VEIN — Mineral body, usually of a vertical or near-vertical nature, striking through the rock for great distances and to great depths.

Bibliography

Adams J, *Mines of the Lake District Fells*, Dalesman, Clapham, 1988

Barton D B, *A History of Tin Mining and Smelting in Cornwall*, D B Barton, Truro, 1967

Borlase E T, "Greenside", *The Engineering and Mining Journal LXXXV*, 1908

Borlase W H, "Greenside Mine", *Institute of Mining Engineers XXV*, 1902-03

Borlase W H, "Greenside Mine", *Institute of Mining Engineers VII*, 1894

Bowen D, *Mines and Quarries Acts*, Sweet and Maxwell, London, 1923

Burt R, *The British Lead Mining Industry*, Dyllansow Truran, Redruth, 1984

Burt R, Waite P, Atkinson M, Burnley R, *Mineral Statistics 1845-1913*, University of Exeter, 1981

Challis P J, *Greenside Mine Ore Dressing 1900 — The Dam Disaster 1927*, Northern Mines Research Society

Clough R T, *The Lead Smelting Mills of the Yorkshire Dales and Northern Pennines*, R T Clough, Keighley, 1980

Collingwood W G, *Elizabethan Keswick*, 1912

Connor C, "Greenside", *Mine and Quarry Engineering*, 1951

Connor C, "Greenside", *Mine and Quarry Engineering*, 1955

Crocker G, *The Gunpowder Industry*, Shire Publications, Aylesbury, 1986

Donald M B, *Elizabethan Copper*, Pergamon Press, 1955

Eastwood T, *Future of Non Ferrous Mining in Great Britain and Ireland*, Institute of Mining and Metallurgy, London, 1959

Griffin A R, *The Collier*, Shire Publications, Aylesbury, 1982

Hamilton H, *The English Brass and Copper Industries to 1800*, Frank Cass and Co, London, 1967

Harwood J, *History and Description of the Thirlmere Scheme*

Hofman H O, *The Metallurgy of Lead*, The Scientific Publishing Co, London, 1899

Holland E G, *Coniston Copper —A History*, Cicerone Press, Milnthorpe, 1987

Holmes P, "Electric Traction at Greenside", *Newsletter No.9*, Cumbria Amenity Trust, 1985

Landless V, *Greenside Explored*, Northern Mine Research Society, 1979

Lewis A, *My First Day in Greenside*, 1986

Little E A, *Chronicles of Patterdale*, 1961

Louis H, *The Dressing of Minerals*, Edward Arnold, London, 1909

McFadzean A, "The Alma — Greenside Revisited", *Newsletter No.9*,

Cumbria Amenity Trust, 1985
McFadzean A, *Wythburn Mine and the Lead Miners of Helvellyn*, Red Earth Publications, Ulverston, 1987
Postlethwaite J, *Mines and Mining in the English Lake District*, reprinted Michael Moon, Whitehaven, 1980
Raistrick A, *The Lead Industry of Wensleydale and Swaledale Vol I*, Moorland, Ashbourne, 1975
Raistrick A, *The Lead Industry of Wensleydale and Swaledale Vol II*, Moorland, Buxton, 1975
Richards R H, *Textbook of Ore Dressing, Hill Publishing*, London, 1909
Rowe D J, *Lead Manufacturing in Great Britain*, Croom Helm, London, 1983
Shaw W T, *Mining in the Lake Counties*, Dalesman, Clapham, 1970
Tyler I, *Force Crag: the History of a Lakeland Mine*, Red Earth Publications, Ulverston, 1990
Wainwright A, *A Pictorial Guide to the Eastern Fells*, Westmorland Gazette, Kendal, 1955
"The Records of Patterdale", *Transactions of the Cumberland and Westmorland Antiquarian and Archæological Society*, Kendal
"Cornish Mines", *Metalliferous and associated Minerals*, 1845-1913

Greenside Archive ref WDB, Kendal-Carlisle
Greenside Archive ref 1979/120, Beamish
Greenside Archive ref BUC Barrow-in-Furness
Greenside Archive ref ZLB Northallerton

Index